INTRODUCTORY
ZOOLOGY
LABORATORY GUIDE

SECOND EDITION

Jeffrey W. Tamplin
Louisiana State University

William B. Stickle
Louisiana State University

Joseph P. Woodring
Louisiana State University

Illustrated by J. W. Tamplin

Morton Publishing Company
925 W. Kenyon Avenue, Unit 12
Englewood, Colorado 80110

Dedication

To Mr. Jackie Cooler,
Dr. Ingemar Larson,
and Dr. Ralph Troll,
three of the best zoology teachers
an aspiring student could ever have.

RECYCLED

PAPER

Copyright © 1997 by Morton Publishing Company

Printed in the United States of America

10 9 8 7 6 5 4 3 2 1

ISBN: 0-89582-374-8

Preface

There are many different approaches to presenting an introductory zoology course. Perhaps the sole uniting thread to all of these alternative strategies lies in a cohesive phylogenetic pattern. One of the fundamental doctrines of biology is that nature always succeeds in producing adaptive features; it is these features and their successive advancements upon which animal phylogeny is based. Our approach to zoology in this manual is one which closely follows a phylogenetic theme and examines the functionality of animal structure. The initial exercises of this manual represent a survey of animal diversity and structural complexity. These chapters focus on the behavior of living animals, their adaptations to the environment, and the dissection of representative species. Following the survey portion of the book are a series of physiological exercises that constitute an experimental section. These experiments are designed to impart an understanding of functional morphology and examine the biological and physical factors that dictate the need for such structures.

This manual is designed to accompany a typical 15 week, one semester introductory zoology course. Each exercise is intended to represent one 2 to 3 hour laboratory period. The manual can easily be modified to accompany a 9 week quarter system by utilizing the first 5 exercises in an introduction to invertebrates course, and the final 7 exercises in a vertebrate structure and function course. Laboratory exams and an occasional field trip to a local museum can be spaced at the instructor's will to complete the required laboratory meetings. A correlation chart, listing popular zoology textbooks and the chapters that accompany the exercises in this manual, is provided to facilitate coordination of a lecture course with the laboratory manual. All of the exercises can be accomplished in a single lab meeting, with the exception of the nitrogen excretion experiment which as written can be completed in a period of several weeks, or transformed to a single period comparative lab by obtaining tadpoles in several developmental stages or by analyzing the nitrogenous wastes produced by adult ranid frogs and those formed by a fully aquatic adult frog such as *Xenopus*. The physiology labs are intended to provide experimental data bases that can be written up as formal laboratory reports. Appendices cover necessary information related to animal architecture and development. Additional appendices discuss the principles behind biological variation and the statistical analysis of such variation. The methodology for presenting a formal scientific paper is also included as an appendix.

Both the living and preserved animals which are discussed at length in this manual are available from biological supply companies. Species examined in this book are usually those that are good taxonomic representatives and are often readily available. Accessory demonstrations are an integral part of a zoology laboratory course because they increase the coverage of organismal and structural diversity. A guide, in the form of a checklist, of suggested demonstrations is presented at the end of each exercise. These checklists can be used by the instructor to indicate which ones will be covered in their particular course, or can be utilized by the students when studying or initially completing the lab exercise.

A systematic overview is presented at the onset of each exercise. While many authors relegate this section, often in miniscule print, to the back of the chapter, we have opted to emphasize systematics and use taxonomy as the organizing force of the manual. Thus, a quick scan of the systematic overview of each exercise will prepare the reader for the forthcoming material. This can be an important learning tool for it equips the mind to better store information. The manual is designed to be read while examining a particular specimen. Instructions on behavioral analyses, structural examination, and tips for dissection are given throughout the text. Metric conversions are listed on the inside front cover; these can be used to help the reader convert the more common U.S. units to the metric units discussed in the text. Useful dissection terms are listed on the inside back cover.

The authors hope that the student finds this manual concise, informative, and visually appealing.

The diversity of the animal kingdom is marvelous; its beauty is only increased when an understanding of function accompanies it.

Acknowledgments

The intricate processes of nature have instilled a fascination of animal life and a respect for its complexity in all of the authors. We thank not only those who have provided pertinent information, but also those who initially implanted our love for zoology. John Lynn and the LSU light microscopy facility graciously donated valuable microscope time used to produce the photographs. Tapash Das, Antonio Todaro, John Fleeger, and Richard Roller were kind enough to provide several electron micrographs which improved the book visually.

Contents

7 Tissue Structure and Function 147

8 Vertebrate External Anatomy, Skeleton, and Muscles 165

9 Vertebrate Digestive, Respiratory, Circulatory, and Urogenital Systems ..179

10 Sensory Systems 191

11 Cardiovascular Function: Heartbeat and Blood Pressure 199

12 Metabolic Processes 203

List of Illustrations and Photographs

LABORATORY SAFETY

Science laboratories should be regarded as a place of learning and inquisition. Ideally, laboratories should harbor a serious atmosphere while still encouraging the students to enjoy themselves. After all, learning can and should be a pleasant and rewarding experience. In any case, students should recognize that a number of potential safety hazards may be present in a laboratory situation. In a typical zoology laboratory most of these hazards are related to electric circuits and equipment, hazardous chemicals, and the presence of potential pathogens, such as bacteria, protozoans, and infectious worms. In some cases, the eggs of nematodes are known to remain viable after exposure to preservatives.

Most of the potential hazards in a zoology laboratory can be avoided by the use of common sense. However, here are a few suggestions to help make the laboratory a safe environment:

1. Always wash your hands after handling either live or preserved specimens; never place objects (pencils, pens, fingers, etc.) into your mouth during the lab period.

2. Avoid eating and drinking in the laboratory. Smoking should also be prohibited as some preservatives are flammable.

3. Never use electrical equipment in areas where there is a chance of exposure to water.

4. If exposure to hazardous chemicals is necessary, wear protective coats, shoes, and goggles.

5. Be extremely careful with scalpel blades. New blades are extremely sharp and can slice through tissue with very little pressure; used scalpel blades may harbor pathogenic infectious agents.

6. Avoid breathing fumes given off by preservatives when dissecting animals. These vapors may cause headaches, eye irritation, or worse.

Systematics is an inherently flexible science. Classification schemes are being continually revised based on new data and new interpretations of data. What one scientist regards as dogma another may dismiss as dog food. This does not limit the usefulness of taxonomy, but merely qualifies its nature and verifies the plasticity of systematics. In this and the following chapters a comprehensive systematic overview is presented that lists major taxonomic categories, their characteristics, and a few representative genera. The schemes we present are currently accepted by a large number of zoologists and are useful in comparative studies of the animal kingdom. Keep in mind that the perspective on these categories occasionally changes and that other authors may advocate slightly different relationships.

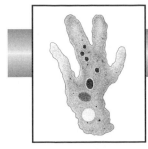

Protozoans, Poriferans, Cnidarians

THE PROTOZOA

SYSTEMATIC OVERVIEW

Unicellular organisms ("protozoa") have traditionally been divided into four major groups based primarily on differences in locomotor organelles and mode of reproduction: the amoebas, the flagellates, the sporeformers, and the ciliates. More recent evidence has indicated that this classification crossed natural evolutionary boundaries, and the scheme that follows places the 65,000 species (30,000 extinct) of protozoans in a more appropriate phylogenetic arrangement. This system recognizes seven separate phyla of unicellular organisms with a variety of primitive plant, fungal, and animal characteristics. We list only four phyla, those with chiefly animal-like characteristics:

PHYLUM SARCOMASTIGOPHORA

Locomotor organelles of flagella, pseudopodia, or both; usually only one type of nucleus and no spore formation; autotrophic, heterotrophic, or both; reproduction usually asexual by fission; body naked or with external or internal skeletons; free living or parasitic (*Amoeba*, *Trypanosoma*, *Volvox*).

PHYLUM APICOMPLEXA

Characteristic set of organelles (apical complex) at anterior end in some stages of the life cycle; spores formed during reproduction; all species endoparasitic; movement by passive diffusion through body fluids or tissue of host; cilia and flagella absent (*Plasmodium*).

PHYLUM CILIOPHORA

Cilia or ciliary organelles present in at least one stage of life cycle; two types of nuclei usually present; reproduction by binary fission although budding and multiple fission (schizogony) also occur; genetic exchange occurs through conjugation; heterotrophic; free living or symbiotic; cell surface usually covered by a pellicle (*Paramecium*).

PHYLUM MICROSPORA

Obligate intracellular parasites which lack mitochondria; minute spores contain a germinal sporoplasm that is injected into a host cell through a coiled tube (*Nosema*, *Thelohania*).

Protozoa are often thought of as primitive, simplified animals. However, no life form is truly simple. Even the most structurally austere organisms execute complex biochemical interactions which allow them to survive in their environment. Protozoa carry on many of the functions of higher animals and are extremely efficient in performing these functions. Each is very well adapted to its environment.

Individual protozoa consist only of a single-cell, although colonial forms exist which show a division of labor rivaling that of multicellular life. As a group they are classified as unicellular protists and placed in kingdom Protista. The kingdom Protista is probably not a natural, cohesive evolutionary grouping, and the organisms placed within show great diversity and incorporate a wide variety of life strategies. Although some subsets within the kingdom are obviously closely related, many of the species placed in the Protista share little in common except their unicellularity.

Protozoans are limited by the very nature of their unicellular form. Every protozoan species, and hence individual, must perform biological functions within the limits of a single plasma membrane. Unicellularity inherently prohibits the formation of specialized organs or tissues, yet the cell is compartmentalized and there is a division of labor within the cytoplasm. This results in the formation of intracellular membrane-bound **organelles** specialized to carry out specific tasks, as well as to separate incompatible biochemical reactions. These organelles provide sensory, locomotive, skeletal, defensive, and contractile functions, among others, for the cell.

They represent a great variety of mostly microscopic (0.005–5 mm) organisms which are widespread ecologically. They are common in marine environments, freshwater, and brackish water environments and some species even live in damp terrestrial habitats. Although many can survive periods of desiccation by producing a dormant cyst or by reproducing sexually (which usually involves a dormant stage), protozoa tend to grow and reproduce only in wet habitats or during moist conditions. While some are free living, others live as parasites (approximately 10,000 species) or in a variety of mutualistic or commensalistic symbiotic relationships.

PHYLUM SARCOMASTIGOPHORA

Subphylum Sarcodina: *Amoeba*

The term "amoeba" is often used as a common name for any of the naked or shelled sarcodines (Figure 1.1). When referring to a specific genus, such as *Amoeba,* the name is capitalized and italicized (or underlined).

As a group, they may either be naked with shifting form or enclosed in a rigid shell (termed a **test**). The test is composed of hardened protein secretions and imparts form and shape to the organism and serves as a primitive exoskeleton. The naked amoebas live in freshwater and marine habitats or among the water droplets of moist soil. The aquatic forms are bottom dwellers and must have a substrate upon which to move, so they are commonly found on aquatic vegetation, submerged wood, or underwater rocks and ledges. The active feeding stage of the life cycle is termed a **trophozoite** and reproduces by **asexual fission**. A dormant **cyst** usually occurs in the life cycle during periods of harsh environmental conditions. Most amoebas feed on algae, protozoans (including other amoebas), bacteria, or rotifers by engulfing the organisms whole. *Amoeba proteus* (Figure 1.2) is a freshwater species that is usually found in slow-moving or still-water ponds of generally clear water.

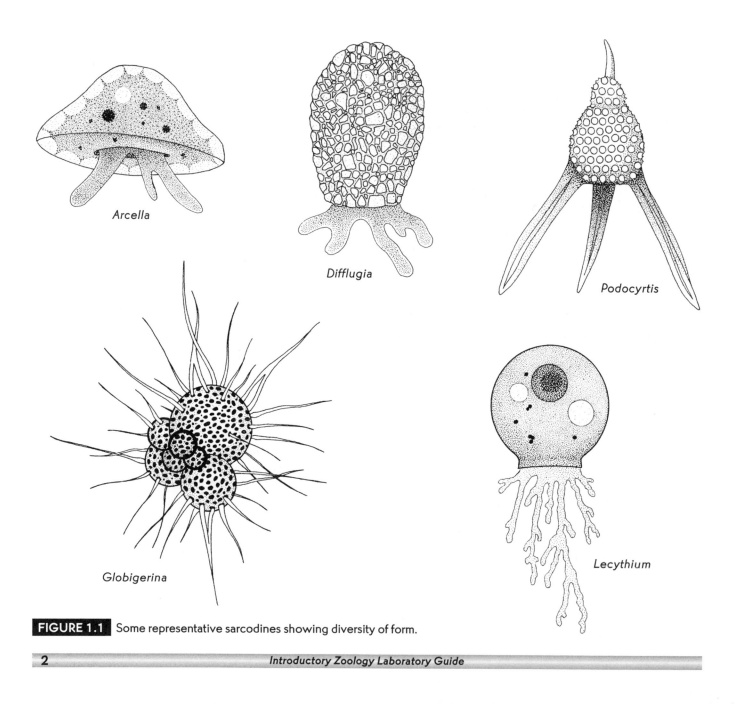

Arcella

Difflugia

Podocyrtis

Globigerina

Lecythium

FIGURE 1.1 Some representative sarcodines showing diversity of form.

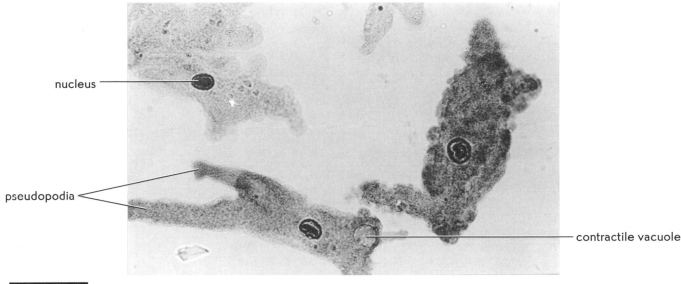

nucleus

pseudopodia

contractile vacuole

FIGURE 1.2 *Amoeba proteus.* Note the characteristic features from Figure 1.3.

PREPARED SLIDES

General structure. Locate the following features on a microscope slide (refer to Figure 1.3). The outermost cell membrane is the **plasmalemma**. The enclosed cytoplasm is differentiated into a thin, clear superficial layer of viscous **ectoplasm** and an inner, fluid granular portion, which is termed the **endoplasm.**

Locomotion. Amoebas propel themselves and change shape by forming temporary cytoplasmic projections termed **pseudopodia.** As each new pseudopodium begins to form, a thickened extension of ectoplasm called the **hyaline cap** forms, and the fluid endoplasm flows toward it. As the endoplasm flows into the hyaline cap, it spills out to the periphery and is converted to ectoplasm, thus building up and lengthening the sides of the pseudopodium like a tube or passageway. At some point the tube is anchored to the substrate by the plasmalemma, and pulls the animal forward. As the tube lengthens, the ectoplasm at the temporary "tail end" converts again to streaming endoplasm to replenish the forward flow. At any time the action can be reversed, the endoplasm streaming back, the tube shortening, and another pseudopodium forming elsewhere and pulling the organism in a different direction.

The cause of the motive force of pseudopodia, and where the motive force is applied is still not completely understood. The classical theory that cytoplasm itself can generate a motive force has recently been reaccepted with the revelation that amoebas possess **microfilaments** similar in structure to the contractile protein filaments of animal muscle tissue. This implies that the contraction mechanism in specialized muscle cells of animal tissue is an adaptation of ancient proteins to a new process and suggests an evolutionary link between protozoans and animals.

Feeding. Amoebas engulf their food whole by a process termed **phagocytosis.** When feeding, amoebas encircle the prey item with pseudopodia and enclose the item along with some fluid from the surrounding medium in a temporary internal membrane. This structure becomes a **food vacuole.** Once an organism has been placed into a food vacuole, digestive enzymes diffuse into these vacuoles and chemically break down the prey. Digestion takes anywhere from 15 to 30 hours depending upon the temperature and other physical factors. Undigested end products can be expelled from the amoeba at any point along the plasmalemma. Amoebas can survive for several days without ingesting food but may decrease cellular volume during this time. Amoebas can also take in fluid droplets and minute nutrient particles by a similar process termed **pinocytosis.**

Nucleus. Locate the genetic storehouse of the organism, the **nucleus.** It is usually oval or disc-shaped, often indented in side view, and finely granulated. The nucleus contains the genetic material of the cell and provides the organism with all the information it needs to grow and reproduce. The nucleus is usually transported along with the cytoplasm as the animal moves and often remains near the center of the cell.

Osmoregulation. One or more water-expulsion vesicles termed **contractile vacuoles** are dispersed through the body of freshwater amoebas. These structures are designed to remove excess water from the cytoplasm. Because the cytoplasm of freshwater amoebas is hyperosmotic to the environment, water tends to diffuse into the cell and must be periodically eliminated to prevent swelling and bursting of the cell. Brackish water amoebas possess contractile vacuoles which are inactive in highly saline waters, but are activated to collect and pump water out of the cell when the environment is dilute. Marine and parasitic amoebas are generally isosmotic to their environment and maintain intracellular concentrations of salts from their own metabolism at levels equal to the osmotic concentration of their environment.

LIVING SPECIMENS

Place a drop of *Amoeba* culture water on a microscope slide. Be sure to scrape the bottom and sides of the container when removing the culture water to facilitate the capturing of several *Amoeba*. On low power scan the slide looking for grayish, irregular shaped, granular organisms. When one is located, switch to high power and observe the movements of a specimen as it creates and forms new pseudopodia. Note the flow of granular cytoplasm through the cell. Place a drop of *Chlamydomonas* or *Paramecium* culture water on the slide near an *Amoeba* and observe the reaction. Scan the slide in search of feeding events.

Subphylum Mastigophora: *Volvox* and *Trypanosoma*

The mastigophorans, or flagellates, are primitive organisms that are considered to be some of the least advanced unicellular life forms. Photosynthetic organisms are characterized by the ability to produce their own organic compounds, utilizing only inorganic environmental sources such as gaseous carbon dioxide, sunlight energy, and water. Animals, on the other hand, are **heterotrophic** and must rely on presynthesized organic compounds for food. This relationship implies that animals depend on **autotrophs**, plants and some bacteria, for their nutrition sources. Many flagellates combine these qualities and display feeding characteristics of both plants and animals. They are considered primitive because they possess one or more long, whip-like flagella, a common (ancestral) characteristic displayed in many plant and animal lineages.

Volvox

Volvox consists of comparatively large (2–3 mm), green, photosynthetic spherical colonies (Figure 1.4).

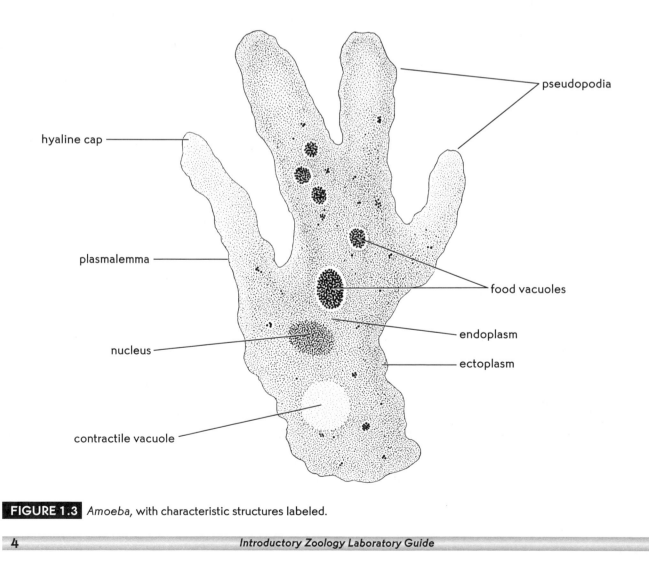

FIGURE 1.3 *Amoeba*, with characteristic structures labeled.

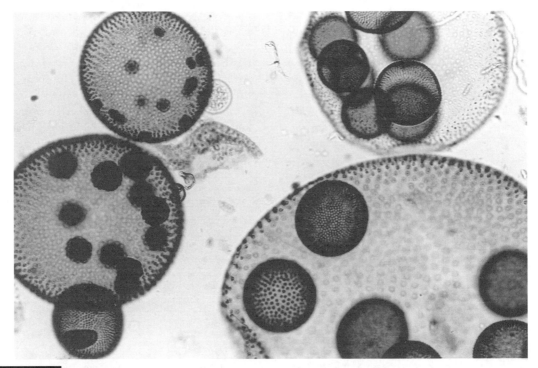

FIGURE 1.4 Several colonies of *Volvox*, a flagellate, showing asexually produced daughter colonies.

Such complex colonies may represent a transition between protozoan and multicellular life forms, and the beginnings of the division of labor (somatic and reproductive cells) and sexual differentiation (sperm and egg cells) are seen in this organism. *Volvox* is common in shallow, freshwater habitats, and is one of the few protists visible with the unaided eye. The colony forms as individual cells remain attached after cell division, and each *Volvox* colony may contain up to 50,000 cells. Volvox reproduces both sexually and asexually; usually sexual reproduction is a contingency mode which occurs when harsh conditions arrive. The resulting zygotes remain dormant until favorable conditions return. Asexual reproduction results in **daughter colony** formation inside the parent body (Figure 1.5).

When these daughter colonies mature they secrete an enzyme that breaks down the gelatinous matrix of the parent colony and releases the daughter colonies into the environment. Observe the movements of a living *Volvox* colony, and note the ultrastructure of colony on a prepared, stained slide.

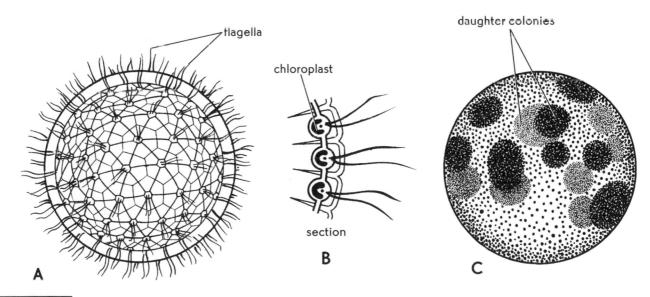

FIGURE 1.5 *Volvox.* A) diagrammatic illustration of a colony showing the relationship between cells; B) section of a colony showing several individual cells, each biflagellated with an eyespot and a cup-shaped chloroplast; C) parent colony showing several daughter colonies.

Trypanosoma

Members of the genus *Trypanosoma* are parasitic flagellates that live in the blood and tissue fluids of nearly all classes of vertebrates. They usually develop in a blood-sucking invertebrate such as an insect or leech, and are transmitted to their vertebrate hosts through the bite of the infected invertebrate. Trypanosomes are common in both the old world and the neotropics and are the cause of devastating diseases. *Trypanosoma brucei* is transmitted by the tsetse fly (*Glossina* spp.) and causes African sleeping sickness in humans. This disease is characterized by fever, weakness, and tremors, and if untreated, ends in prolonged coma and finally death. A related disease, "Nagana," kills cattle and other domestic animals and has hampered economic development in many regions of Africa. *Trypanosoma cruzi* is common in South and Central America and causes Chagas' disease in humans. It has symptoms similar to those of African sleeping sickness and is transmitted through the bite of the "kissing bug."

Study the slides of stained blood smears infected with trypanosomes and notice the features identified in Figure 1.6. The trypanosomes appear as small

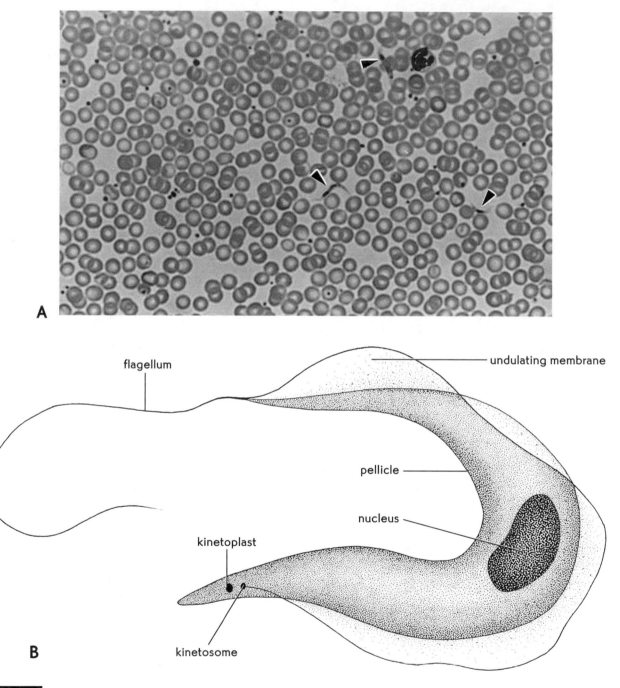

FIGURE 1.6 *Trypanosoma*, a parasitic protozoan which is carried by blood-feeding invertebrates and causes disease in vertebrate circulatory systems. A) blood smear photomicrograph showing several trypanosomes circulating among the plasma and corpuscles; B) illustration depicting structure.

sickle-shaped organisms in the plasma circulating among the blood corpuscles. Note the fusiform, slightly twisted body of the organism. The shape of the organism is imparted by a stiff **pellicle,** an external protective sheath composed of one or more cell membranes reinforced by underlying **microtubules.** There is an anterior flagellum and along one side of the organism is an **undulating membrane.** The flagellum originates posteriorly, travels along the surface of the body close to the pellicle, and continues anteriorly as a free whip. When the flagellum beats, the pellicle is pulled up into a fold; the fold and the flagellum together constitute the undulating membrane.

PHYLUM APICOMPLEXA: *PLASMODIUM*

The apicomplexans or "sporeformers" are entirely endoparasitic. The phylum is named for a complex of organelles located at the anterior end, or "apex," of the cell. The apical complex allows the organism to penetrate through the tissue layers or cell membranes of the host, so consequently, the apical complex is often present only during the stages of the life cycle in which host penetration occurs. Because members of this phylum lack special locomotor organelles, movement is typically accomplished by passive diffusion through the body fluids of the host, although some can move by gliding or by changing their body shape. Some species have flagellated gametes. They reproduce both sexually and asexually and the life cycle often requires an **intermediate host.** The intermediate host is usually an invertebrate (often a mosquito) and the **final,** or **definitive, host** is generally a vertebrate. At some point in the life cycle, a resistant "spore" is formed which is capable of infecting the next host. The process of **sporogony,** or spore formation, in most apicomplexans follows the formation of the zygote, so that sexual reproduction results in the production of spores.

As a mechanism to increase population numbers, and hence the odds of infecting the next host, many sporeformers also undergo asexual multiple fission, or **schizogony,** which leads to great numbers of **sporozoites** being formed. Apicomplexans have no mouth, and nutrients are absorbed directly from the surrounding environment. The most notorious apicomplexan is *Plasmodium,* an organism which causes "Malaria" in humans. There are over 100 known species of *Plasmodium,* of which 4 are pathogenic to man. The remaining species parasitize a variety of terrestrial vertebrates.

Nearly 200 million people are infected with Malaria each year, mostly in tropical regions. Approximately 1 million people die each year from this disease in Africa alone. Observe a stained slide of *Plasmodium.* Note the stage shown and region of infection and locate it in the following life cycle. Major stages and regions of infection are identified in bold print.

Life Cycle

Plasmodium is transmitted to the human by a female *Anopheles* mosquito that has previously had a blood meal from a malaria-infected human. The mosquito, while feeding, injects into the **human bloodstream** some of its saliva, which contains infective **sporozoites.** Once inside the bloodstream, the sporozoites migrate to the **liver** and penetrate individual liver cells. This is the **incubation** period and can last from 6 to 15 days depending on the species. Inside the liver cells, the sporozoites undergo asexual **schizogony** producing **merozoite** offspring. The merozoites break out of the liver cells and reenter the bloodstream, attacking and penetrating red blood cells (**erythrocytes**). Further schizogony of the merozoites in the red blood cells increases their population, and the merozoites break out of the red blood cells cyclically, every 48 to 72 hours, again depending upon the species.

As these merozoites exit the red blood cells they release their metabolic **toxins** and this causes repeated outbreaks of malarial symptoms. Typical symptoms are fever, chills, and sweating. The synchronous release of these toxins may lead to death for the weakened victim. The released merozoites may migrate back to the liver or they may continue to cycle and reproduce among the red blood cells.

After several phases of schizogony, some merozoites develop into sexual **gametocytes.** These gametocytes float freely in the bloodstream and may be ingested by a mosquito while biting an infected vertebrate. When the gametocytes reach the mosquito's **gut,** they mature into sex cells, or **gametes.** If male and female gametes encounter each other in the gut, they fuse (in **fertilization**) and create a **zygote.** The zygote penetrates through the tissue of the mosquito's **stomach** and transforms into the spore stage, termed an **oocyst.** Inside the oocyst, **sporogony** occurs forming thousands of sporozoites. These sporozoites migrate to the **salivary gland** of the mosquito and are ready to infect a new host should the mosquito take a new blood meal. (See Figure 1.7)

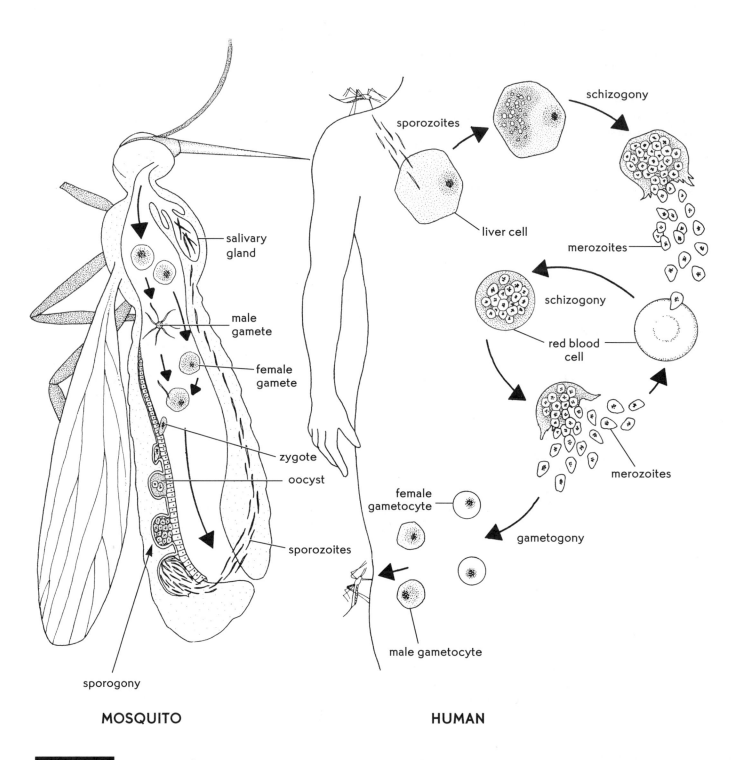

MOSQUITO   **HUMAN**

FIGURE 1.7 *Plasmodium* life cycle. *Plasmodium* causes the disease "Malaria" in man and requires two hosts to complete the life cycle. The left side of the illustration depicts the stages present in the intermediate host (a mosquito), and the right side shows the stages which infect the definitive host (humans).

PHYLUM CILIOPHORA: *PARAMECIUM*

The phylum Ciliophora is the most diverse and successful of all the major protozoan groups. They occupy a variety of aquatic habitats and are known for their agility and maneuverability. Their primary adaptive feature is the possession of **cilia**, small hair-like locomotory projections which extend from the cell. *Paramecium* (Figure 1.8) is an active, highly maneuverable ciliated protozoan common in most freshwater environments that contain vegetation and decayed organic matter. *Paramecium* feeds on bacteria, algae, and smaller protozoa which are common in its environment and displays many features characteristic of the phylum: **cilia**, **macro-** and **micronuclei**, **trichocysts**, and a compound **pellicle**.

LIVING SPECIMENS

Locomotion. Place a drop of *Paramecium* culture and add a drop of "Proto-slo" or 10% methylcellulose (a media thickening solution) on a slide and spread it evenly. Leave the slide uncovered and examine the medium with the scanning (low power) lens of a compound microscope to locate the specimen. You may see numerous paramecia swimming in seemingly random patterns. Actually they are processing environmental stimuli and reacting just as any higher organism would. Locate a single *Paramecium* that is still, rotate the lens to high power and observe the structure of the organism.

The *Paramecium* is elongate and slipper-shaped, somewhat transparent and colorless. Some species are quite large, comparatively, but most species average 10 micrometers in length. Paramecia can swim at the rate of 1,000 micrometers per second, a rate equivalent to 409 miles per hour for a 6 foot human being. It tends to spin on its axis and move somewhat erratically in a particular direction. The *Paramecium* displays a characteristic "avoidance reaction" that consists of a series of 3 negative responses: rapid reverse; spinning; and forward movement in an alternate direction. This reaction occurs when normal forward progress is impeded by mechanical obstructions, extreme temperature changes, or by critical levels of certain chemicals.

General structure and function. From your specimen and Figure 1.8, note the **oral groove** that extends obliquely from the anterior end to the middle of the body. At the posterior end of the groove is the **mouth** (**cytostome**). From the mouth, the oral groove enters the body as a small canal. The groove and **cytopharynx** are lined with strong cilia that are used in the capture of food. The cytoplasm consists of two zones, the clear outer **ectoplasm** and the inner **endoplasm**. Outside the ectoplasm is a shape-imparting **pellicle**. A delicate **plasma membrane** lies just beneath the pellicle. Both of these structures are better seen in stained preparations.

Osmoregulation. A **contractile vacuole** (water-expulsion vesicle) is usually located in both the anterior and posterior ends of the body. The vacuoles alternately fill and empty, depending on the physiological activities of the organism. After one empties, star-like **radiating canals** are often visible. These radiating canals collect liquid from a network of minute tubules and empty into the vacuoles which, when filled, rupture, and expel the fluid outside of the organism's cell.

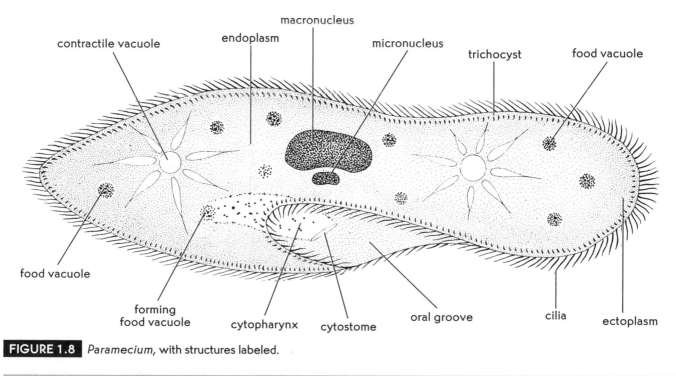

FIGURE 1.8 *Paramecium*, with structures labeled.

The vacuoles and radiating canals are most prominent in animals that have settled down, when most of the culture water on the slide has evaporated. The pulsating of the vacuoles depends on the temperature and osmotic pressure of the surrounding fluid. As the temperature increases, the pulsation period decreases. The pulsation rate can be slowed down by adding small crystals of sodium chloride (table salt) to the medium.

Cilia. These fine hair-like projections act in a manner similar to the oars of a boat; in other words, they have an effective stroke that propels the animal forward and a recovery stroke that offers little resistance. Cilia are typically shorter than flagella, but the individual infrastructure and action of the two are similar. Both cilia and flagella have a 9+2 arrangement of microtubule doublets which slide past each other laterally during undulation. A single pair of microtubules is surrounded by nine peripheral pairs of microtubule doublets. The specific arrangement of the cilia are adaptations of each species to its particular lifestyle. Some ciliophorans scurry about on "legs" constructed from cilia clusters bonded together; other species have tightly-packed rows of cilia which are synchronized for fluid undulation. Some species of *Paramecium* possess 10,000 to 14,000 cilia. Because the cilia lie so closely together, fluid movement produced by one cilium triggers the beating of adjacent cilia, resulting in patterns of coordinated movement.

Ciliates are generally adapted for speed and maneuverability. The propulsive force of thousands of tiny coordinated cilia is much greater than that of a few long flagella, a factor in the evolutionary success of ciliates, which are usually faster, larger, and more diverse than flagellates.

Nuclei. Ciliates possess two different types of nuclei. The **macronucleus** is **polyploid** and has multiple copies of the diploid number. Macronuclei genes are not distributed in chromosomes but are packaged in large numbers of small units — each unit having many copies of only a few genes. The macronucleus divides mitotically, and controls physiological, somatic operations of the cell. The **micronucleus** is **diploid**, divides meiotically, and is utilized genetically in reproduction. The number of nuclei in various ciliates may vary from one of each kind to several of either kind. Often the micronucleus is present in large numbers. Some species of *Paramecium* have as many as 80 micronuclei per cell.

Feeding. *Paramecium* is a particulate feeder; it lives on small particles such as bacteria and algae which it moves toward the cytostome by the action of cilia in the oral groove. Place some bacteria or single-celled yeast into the medium and attempt to feed the organism. As your specimen slows down, note the currents created by the cilia in its **oral groove**. Watch the passage of the food particles into the groove and through the **cytostome**, or cellular mouth, into the **cytopharynx**. Observe the formation of a **food vacuole**, which is a membrane-bound sac containing water and suspended food particles. Food vacuoles circulate through the cytoplasm in a distinct pattern. This phenomenon is termed **cyclosis**. The food is digested by enzymes secreted into the food vacuole, and the indigestible remains are expelled from the body through an opening in the pellicle termed the **cytoproct**.

PREPARED SLIDES

General structure. Examine a *Paramecium* on a stained slide using both low and high power objective lenses. Its entire surface is covered with **cilia**, but these may be difficult to see with a light microscope. Examine the **pellicle**. Embedded in the ectoplasm and penetrating the pellicle are small weapon-like structures termed **trichocysts**. Trichocysts expel long filamentous structures with spear-like tips used to deter predators. Identify the **macro-** and **micronuclei**. Note whether there are multiple copies of each. See if you can find any **food** or **contractile vacuoles**.

Binary fission. Study a stained slide of paramecia undergoing **binary fission** (Figure 1.9). Fission is the simplest type of asexual reproduction and usually occurs when environmental conditions are favorable. Various stages are probably present, but *Paramecium* fission is characterized by posterior separation, so that the **daughter organisms** pull apart end-to-end. The macro- and micronuclei separate individually and complementary structures end up in both halves just before complete separation.

Conjugation. Observe a stained slide of **conjugation** (Figure 1.9). Conjugation is a specialized type of sexual reproduction because genetic exchange occurs without the aid of gametes. Many paramecia will be present, but those undergoing conjugation will be joined longitudinally and lying with their oral grooves attached. A **conjugation tube** is formed between separate individuals which facilitates the exchange of micronuclear material. Asexual fission follows conjugation resulting in daughter cells with new genetic combinations.

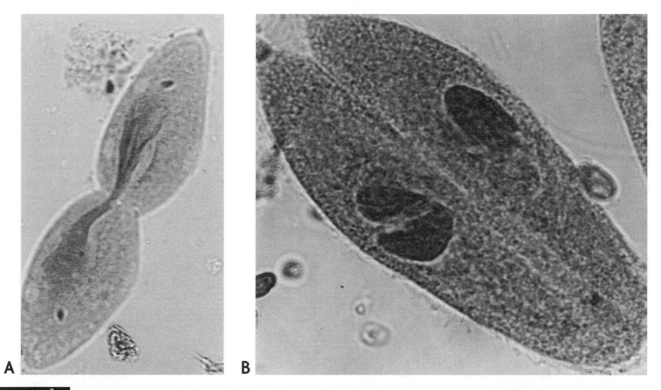

FIGURE 1.9 *Paramecium* reproduction. A) binary fission; B) conjugation between two individuals lying with oral grooves attached.

PHYLUM MICROSPORA

Microsporidians are an enigmatic group of intracellular parasites which are capable of infecting a variety of animals, primarily insects (Figure 1.10). Most individuals are species-specific, only infecting a particular host. Microsporidians produce tiny (5 µm) spores which are the infective stage of the life cycle. The spores are injected into host cells through coiled tube-like structures. Unique in comparison to other eukaryotic cells, microsporidians lack mitochondria and probably obtain the energy they need for growth from their hosts.

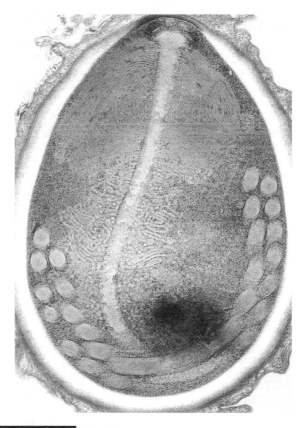

FIGURE 1.10 *Thelohania*, a microsporidian. This is the spore stage of an undescribed microsporidian intracellular parasite, found in the muscle tissue of the blue crab, *Callinectes sapidus*.

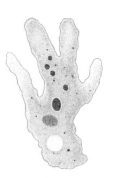

✓ Checklist of Suggested Demonstrations

THE PROTOZOA

_____ 1. *Amoeba proteus* whole-mount slide with accompanying charts and diagrams.

_____ 2. *Amoeba proteus* living specimens in culture.

_____ 3. *Volvox* whole-mount slide with accompanying light micrographs and diagrams.

_____ 4. *Volvox* life cycle diagram.

_____ 5. *Volvox* living colonies in culture.

_____ 6. *Trypanosoma* blood smear slide with accompanying diagram.

_____ 7. *Plasmodium* blood smear slide.

_____ 8. *Paramecium caudatum* whole-mount slide and accompanying diagram.

_____ 9. *Paramecium* fission slide.

_____ 10. *Paramecium* conjugation slide.

_____ 11. *Paramecium* living specimens in culture.

Protozoan Notes and Drawings

Protozoan Notes and Drawings

THE SPONGES (PORIFERANS)

PHYLUM PORIFERA

The members of the phylum Porifera (Figure 1.11) are among the simplest metazoans in the animal kingdom. The ancient Greeks and many following generations believed that sponges were actually a form of plant life. Adult sponges are all **sessile** and many have no predetermined body architecture (**asymmetrical**) and grow strictly in response to environmental conditions. Some advanced forms have a characteristic circular shape and are **radially symmetrical.** Sponges are little more than a loose aggregation of cells, which show little or no tissue organization, hence they are restricted to the **cellular level of organization.** There is a division of labor among cells, but there are no true organs, certainly no organ systems, not even a mouth or digestive tract, and only extremely primitive nervous communication is possible between cells. There are no embryonic germ layers; the adult body consists only of inner and outer epithelial layers intertwined about a series of canal systems. Sponges may be either solitary or colonial.

Their primary adaptive features are their **pores** and **canal systems,** the flagellated **choanocytes,**

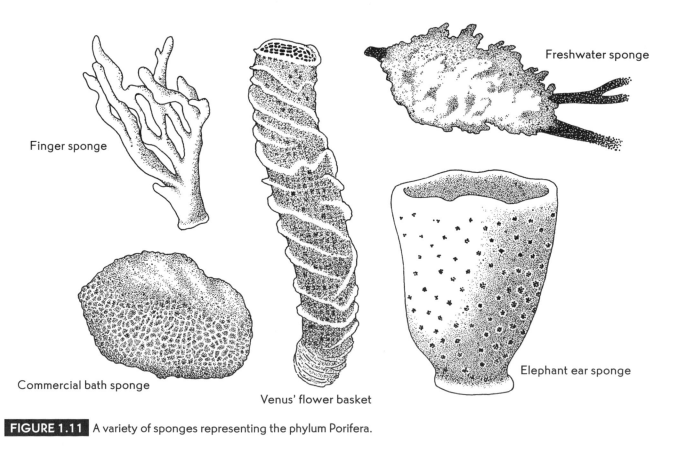

Finger sponge

Commercial bath sponge

Venus' flower basket

Freshwater sponge

Elephant ear sponge

FIGURE 1.11 A variety of sponges representing the phylum Porifera.

that line their cavities and create currents of water, and their unique internal skeletons of **spicules.** Many species have an internal cavity, termed a **spongocoel,** that opens to the external environment through a central opening (an **osculum**). Most sponges are marine, but there are approximately 150 freshwater species that are found attached to sticks, leaves, or other objects in ponds and streams of low current velocity. Sponges range in size from a few millimeters to over 1 meter in height. Body plans range from a simple canal system with a large central spongocoel to arrangements with intricate water passages which lack a central cavity.

Sponges are sedentary filter feeders and can trap food particles (mainly bacteria) suspended in the water which range in size from 0.1 to 50 μm. These particles consist of organic material and small planktonic organisms. All digestion is intracellular and accomplished by **food vacuoles.**

General structure and function. In their most basic form sponges are little more than a hollow tube, closed at one end and open at the opposing end. The wall of the tube consists of two layers of cells separated by a thin noncellular layer, the **mesohyl.** The outer layer of cells is termed the **pinacoderm.** It consists of flattened cells (**pinacocytes**) that serve as an epithelial layer. The inner layer is composed of flagellated **choanocytes** ("collar

cells") that create the water current inside a sponge and trap food particles. Closely associated with the choanocytes, are digestive **amoebocytes.** Amoebocytes transport digested nutrients to other body cells and perform a variety of important life functions.

Skeletal elements. The support system of sponges consists of a unique meshwork of needle-like **spicules** (Figure 1.12A). These spicules differ in shape, size, and material and it is these differences which are the primary taxonomic tool used in classifying sponge species. Spicules may be composed of **calcium carbonate (limestone; $CaCO_3$), silica dioxide (SiO_2),** or protein fibers termed **spongin.** Some species have a combination of silica spicules embedded in a network of spongin fibers. The four classes of sponges are divided primarily on the type of spicule material present.

Sponge reproduction. Both asexual and sexual reproduction occur within the sponges. Asexual reproduction often occurs as **budding** or in freshwater species as **gemmule** formation. Gemmules (Figure 1.12B) are small reproductive bodies with a thick layer of dead protective cells and spicules surrounding young, living cells. Gemmules can be produced year-round but are often produced in response to harsh environmental conditions and are often dormant until favorable conditions return.

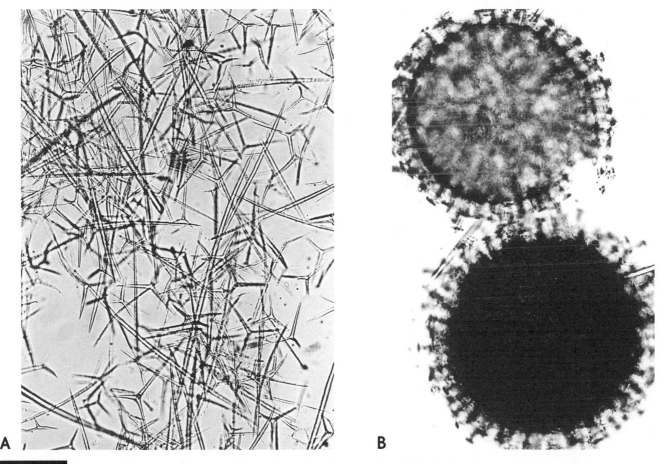

A   **B**

FIGURE 1.12 Sponge structures. A) an assemblage of spicules; B) gemmules of the freshwater sponge, *Spongilla.*

Sexual reproduction usually occurs when environmental conditions are optimal and the sponge is growing. Most sponge species are bisexual (or **hermaphroditic**) and are capable of producing both male and female gametes, however the two types are usually produced at different times to prevent self-fertilization. The sperm cells are released into the water column where they are picked up by neighboring sponge choanocytes. These choanocytes lose their collar after ingesting sperm cells, and enter the mesohyl carrying the sperm to the eggs. The resulting zygotes develop into flagellated larvae, either **parenchymula** larvae or **amphiblastula** larvae, depending upon the type of sponge.

Morphology. Sponges exhibit three distinct body morphologies based on differences in their organization and placement of their internal canal systems (Figure 1.13). The simplest form (**asconoid**) possesses a large

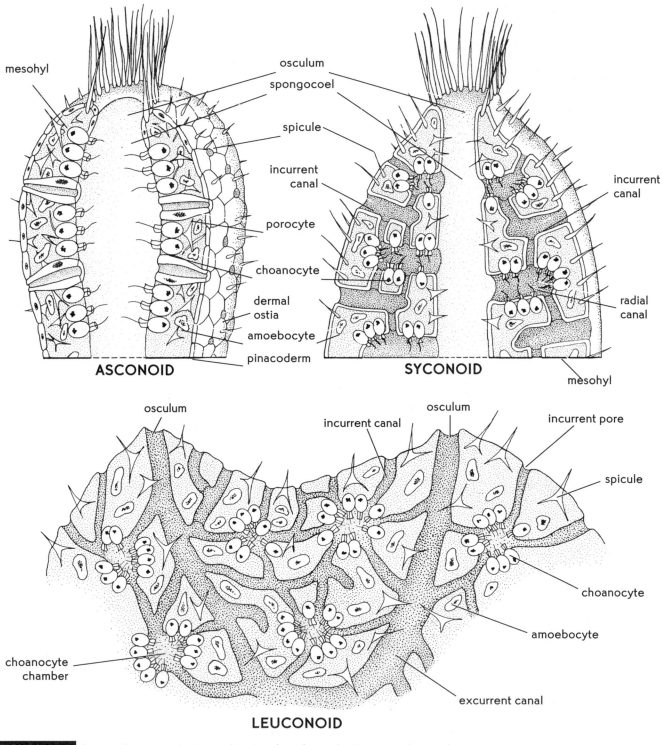

FIGURE 1.13 Sponge body morphologies showing the relationship between the canal structure, choanocyte position, and presence of the spongocoel among asconoid, syconoid, leuconoid sponges.

spongocoel and osculum, a simple linear canal system, and the choanocytes line the spongocoel only. The **syconoid** form is characterized by an elaboration of the canal systems, a thicker body wall, and the presence of pinacocytes lining the spongocoel. **Radial canals** form pouches which project from the central spongocoel and are lined by choanocytes. Between the radial canals lie **incurrent canals**. The openings which connect the radial canals with the spongocoel are termed **apopyles** and the channels which connect the radial canals with the incurrent canals are termed **prosopyles**. The most advanced sponge morphology is the **leuconoid** form, and this type is the most successful and hence most common among the sponges. Leuconoid sponges are typically colonial and lack a large spongocoel. The canal system is complex and highly branched with many internal chambers which are lined with choanocytes. Often the external pores are larger than those of the simpler morphologies and usually more than a single osculum exists.

PREPARED SLIDES

Examine a prepared slide of a cross-section of *Sycon* (= *Granitia* or *Scypha*). The whole organism is vase-shaped, with a body wall made up of a system of tiny, interconnected dead-end canals whose flagellated cells draw in water from the outside through minute pores, remove food particles and oxygen, and then empty it into a large central cavity (spongocoel) for expulsion to the outside. All sponges have some variation of this general theme of canals and pores on which they depend for a constant flow of water.

Note the **spongocoel** in the middle of the section (Figure 1.14). Study the canal system. Find the flagellated chambers (**radial canals**), which open into the spongocoel through the **apopyles**. Since these openings are smaller in diameter than the radial canals, some of them will be lacking in this section, and some of the radial canals will appear closed at the inner end. Identify the **incurrent canals**, which open to the exterior by the **dermal ostia**. Follow these canals inward and note that they also end blindly. Water passes from the incurrent canals into the radial canals through a number of tiny pores, or **prosopyles**, which may not be evident on the slides.

Choanocytes. With high power, observe the "collar cells," or choanocytes (Figure 1.15), that line the radial canals. Although they are flagellated, you probably will not see the flagella. The choanocytes are important in feeding as small particles are trapped by the collar and transported to the amoebocytes for digestion. Choanocytes have three primary features, each of which is adapted to perform a specific feeding function. The

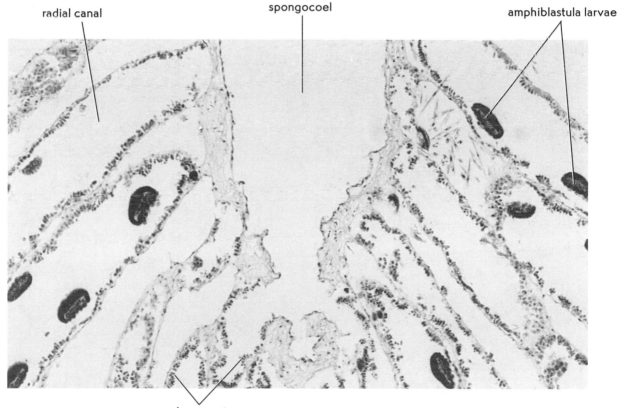

radial canal spongocoel amphiblastula larvae

choanocytes

FIGURE 1.14 Partial cross-section through *Sycon,* a syconoid sponge.

flagellum creates the water current which propels water and nutrients through the sponge. The **collar** surrounds the flagellum and consists of numerous **microvilli** which form a net-like structure capable of trapping food particles. Lying at the base of the collar is the **cell body** which engulfs the food particles trapped by the microvilli, forms food vacuoles, and transfers the vacuoles to the amoebocytes. A typical sponge requires approximately 30 minutes to 1 hour to transfer the food vacuole, and digestion is complete within 4 to 5 hours.

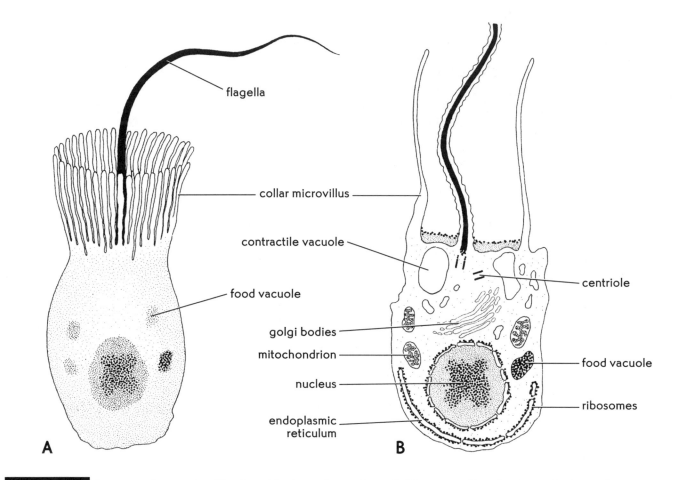

FIGURE 1.15 A sponge choanocyte. A) external features of an entire cell; B) cross-section showing major internal structures.

✓ Checklist of Suggested Demonstrations

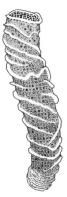

THE PORIFERANS

_____ 1. Miscellaneous dried sponges.

_____ 2. Syconoid sponge cross-section slide.

_____ 3. Sponge anatomy charts and diagrams.

_____ 4. *Spongia* (commercial bath sponge) dried specimen.

_____ 5. *Cliona* (boring sponge) specimen and clam shell with scars.

_____ 6. *Euplectella* (Venus' flower basket) dried specimen.

_____ 7. Sponge spicule strew slide.

_____ 8. Sponge gemmules slide.

_____ 9. *Spongilla* living specimen.

Poriferan Notes and Drawings

Poriferan Notes and Drawings

RADIATE ANIMALS (CNIDARIANS)

SYSTEMATIC OVERVIEW

PHYLUM CNIDARIA

Aquatic, radial or biradial eumetazoans with cnidocytes (stinging cells); 11,000 living species.

Class Hydrozoa

Most have polyp and medusa stages with noncellular mesenchyme; cnidocytes restricted to epidermis; medusa possesses a velum in most species; freshwater and marine, solitary or colonial (*Hydra, Gonionemus, Obelia, Physalia*).

Class Scyphozoa

Marine jellyfish; medusa is the dominant body form and lacks a velum; polyps absent are reduced; mesenchyme is cellular; cnidocytes occur in both epidermis and gastrodermis (*Aurelia*).

Class Anthozoa

Sea anemones, corals, sea pansies, and sea fans; marine polyps with no medusae; well-developed tissues in the mesenchyme; cnidocytes in both epidermis and gastrodermis; body parts arranged in multiples of six or eight (*Astrangia, Gorgonia, Renilla, Tubipora*).

PHYLUM CTENOPHORA

Comb jellies or sea gooseberries; 100 species (*Pleurobrachia*).

which display it. Radial organisms (phyla Cnidaria and Ctenophora, formerly united as the Coelenterata) are uncephalized and radial symmetry is an adaptation for both a sessile, benthic existence, as well as a pelagic lifestyle. This phylum has many unique features which separate them from the rest of the animal kingdom, but the basis of many structural and organizational features common to higher metazoans are first seen here. Similar cells tend to group together and form rudimentary tissues (hence, radial animals represent the **tissue level of organization**). They have, however, only one really well-defined tissue, the **nervous tissue.**

Cnidarians are primarily marine organisms with a few freshwater exceptions. The body is arranged around an oral-aboral (mouth to base) axis with **tentacles** surrounding a central **mouth.** Cnidarians are carnivorous animals, utilizing the tentacles to capture and manipulate their prey. The cnidarian body is composed of three layers: the outer **epidermis;** the median **mesoglea;** and the internal **gastrodermis.** Among different species, the mesoglea ranges from a thin, noncellular layer to a thick, cellular jelly-like mass. Regardless of complexity, the mesoglea is derived from both the epidermis and the gastrodermis, thus cnidarians are **diploblastic** (derived from two embryonic layers, the ectoderm and the endoderm).

Cnidarian characteristics and advancements over lower metazoans include: the presence of embryonic **germ layers** homologous to those of more advanced animals; the development of a digestive, circulatory, and respiratory structure, the **gastrovascular cavity;** the presence of "stinging cells," termed **cnidocytes;** and the presence of a skeletal or support system (which may be either **calcareous** or **hydrostatic**).

The cnidarians show two contrasting body morphologies (Figure 1.16), each of which is adapted to a

PHYLUM CNIDARIA

Radial symmetry is a primitive body architecture, yet it is an adaptive and successful plan for those organisms

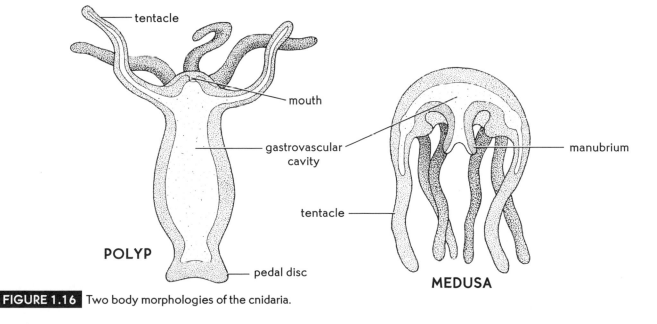

FIGURE 1.16 Two body morphologies of the cnidaria.

different set of conditions and utilizes a different food source. The **polyp** form is designed to attach to a substrate and lead a sessile existence. The **medusa** is a free-swimming (pelagic) body form adapted to float about in the water column. In some groups of cnidarians both polyp and medusa stages are found in the life cycle, usually each form produces the other in an **alternation of generations;** in others, such as the sea anemones and corals, there is no medusoid form; and in the jellyfish, the polyp stage is reduced or lost completely. In life cycles in which both polyps and medusae are found, the juvenile polyp stage gives rise asexually to the medusa, which reproduces sexually. Both the polyp and medusa have the **diploid** number of chromosomes, but the gametes are **haploid.**

Class Hydrozoa: *Hydra* and *Physalia*

Hydra

Hydras (Figure 1.17) are common freshwater animals that live attached to leaves and twigs in ponds, lakes, and streams. They are just visible (2–25 mm in length), and their reactions and feeding behavior can be studied alive, preferably with the aid of a dissecting (binocular) microscope. Hydras are solitary polyp forms and are somewhat atypical of the class because they lack the medusa stage. Also, most hydrozoan polyps are colonial while *Hydra* is a solitary species, and the polyp bears gonads, a characteristic typical of the medusoid form.

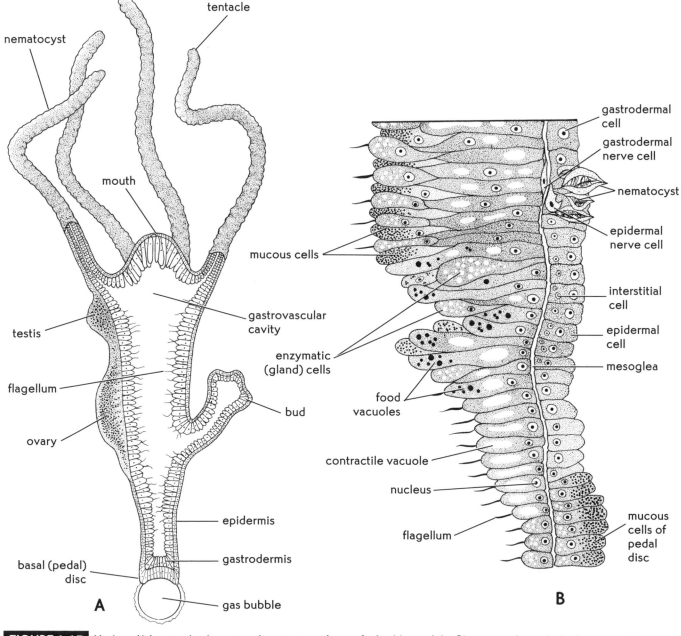

FIGURE 1.17 *Hydra.* A) longitudinal section showing topology of a budding adult; B) section through the body wall showing cell differentiation.

LIVING SPECIMENS

Behavior. Place a living *Hydra* in a watch glass along with a few drops of culture water. The hydra may be contracted at first or extended. Observe its movements in a binocular or dissecting microscope as it recovers from the shock of transportation and begins to examine its new environment. The bumpy appearance of the body, especially the tentacles, is caused by clusters of specialized **cnidocytes**, which contain the stinging capsules termed **nematocysts** (Figure 1.18). After the *Hydra* becomes acclimated to its new surroundings, it will begin to probe and search the environment by extending its tentacles and in some cases by dislodging itself from the substrate. Observe its behavior as you touch it with a needle probe. Alternately increase and decrease the light intensity on the specimen and note if there is a reaction.

Add several *Daphnia* ("water fleas") to the watch glass and observe the interaction of the *Hydra* and *Daphnia*. Even though some *Daphnia* are larger than most *Hydra*, the *Hydra* will capture and immobilize the *Daphnia* with its tentacles and cnidocytes. Watch the *Hydra* as it engulfs the prey item and digestion begins in the gastrovascular cavity. Note that the cavity will darken as it becomes stocked with nutrients.

PREPARED SLIDES

Study a stained slide of both a whole mount and a cross-section of the body. Examine under both low and high power. The body wall is made up of two layers of cells, the outer **epidermis** (formed from embryonic ectoderm) and the inner **gastrodermis** (from the endoderm). These two layers are separated by a thin layer, the **mesoglea**.

Epidermis. In hydrozoans, the bulk of the epidermis consists of medium-sized cuboidal cells with darkly stained nuclei. The inner ends of these **epitheliomuscular** cells constrict into thin fibers (**myonemes**) that run longitudinally in the mesoglea and allow rapid contraction of the body wall. The contractile fibers are an integral part of the **nerve net**, which lies just beneath the epidermal layer. Embedded among the epitheliomuscular cells are several types of specialized cells. **Gland cells** in the epidermis secrete mucus to protect the body from desiccation and those in the mouth region lubricate food items as they pass into the gastrovascular cavity. Gland cells in the pedal disc of polyps help keep the organism attached to substrates. **Cnidocytes** (stinging cells) containing the **nematocyst** capsule are located periodically along the epidermis and used in prey capture and immobilization. They are found in greatest number along the surface of the tentacles. At the base of the epitheliomuscular cells are some small dark **interstitial** cells. These undifferentiated embryonic cells have the capability to develop into many of the specialized cells when required.

Mesoglea. The mesoglea is a thin, noncellular layer secreted by cells in either the epidermis or gastrodermis. It contains a high proportion of collagen fibers that impart an elastic skeletal quality to the body wall. Although noncellular, the mesoglea possesses a few wandering amoebocytes. The contractile or nerve fibers in it may not be visible.

Gastrodermis. The gastrodermis is composed of flagellated, columnar **nutritive-muscular cells**. The flagella create currents in the gastrovascular cavity, circulating food and water. These cells also engulf food items. Contractile fibers at the base of the cells run in a circular direction in the mesoglea. When these fibers contract, the diameter of the hydra decreases, thus the organism becomes longer and thinner. **Gland cells** located in the gastrodermis secrete mucus and digestive enzymes into the gastrovascular cavity. The gastrodermis may appear as two separate layers of cells because the outer part of the cells contain large fluid-filled vacuoles, and the internal ends of the cells contain darker food vacuoles that have been created by phagocytosis. Digestion in all cnidarians is both extracellular and intracellular. Extracellular digestion occurs in the lumen of the gastrovascular cavity. This involves protein hydrolysis as well as some carbohydrate breakdown and softens and prepares the food for phagocytosis. Intracellular digestion follows and occurs in the nutritive-muscular cells.

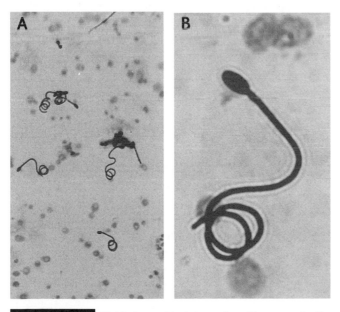

FIGURE 1.18 Cnidarian stinging cells. A) several discharged nematocysts; B) detail of a glutinant capsule.

Asexual reproduction. Asexual reproduction is often associated with the polypoid form and occurs during times of favorable environmental conditions. It can be accomplished either through **budding, fission, or regeneration.** Budding is the most common method employed and is useful in species dispersal. When a bud forms, a hollow outgrowth of the body wall lengthens and develops tentacles and a mouth at its distal end. Eventually the bud constricts at the basal end and breaks off from the parent.

Observe a stained slide of *Hydra* exhibiting bud formation. Note the relationship of the bud to the parent. The gastrovascular cavities of the two are continuous and both layers of the parent wall extend into the bud.

Sexual reproduction. Sexual reproduction usually occurs in the medusoid form and is often triggered by the onset of harsh environmental conditions. Some species of hydrozoans are **monoecious** and have both testes and ovaries on the same individuals; other species are **dioecious** and possess separate sexes. The gonads, or sex organs, are simple clusters of gamete-producing cells that develop from the interstitial cells. The **ovary** is found near the basal end of the organism and produces a large egg, or **ovum.** The **testes** are smaller outgrowths near the oral end which release **spermatozoa.** The sperm swim to the egg and fertilize it in position. The resulting **zygote** (fertilized egg) develops into a solid **blastula** and then into a ciliated **planula** larva. Planula are free-swimming and feed on microorganisms or are nourished by a yolk. After a short period of time, they attach to a substrate and develop into polyps.

Physalia

The Portuguese man-of-war (Figure 1.19) is an extremely complex, colonial hydrozoan which simultaneously displays both polyps and medusae. Because *Physalia* is a pelagic marine organism with hanging tentacles, it is often mistaken for a scyphozoan jellyfish. The entire colony may contain up to 1,000 individuals. The polyps of the colony are specialized for feeding, asexual reproduction, and prey capture. The medusoids are adapted for locomotion, flotation, and sexual reproduction. The polypoid individuals include **gastrozooids** which are specialized for feeding and digestion, **dactylozooids** (fishing tentacles) used in prey capture, and **gonozooids** associated with reproduction. The medusae include dioecious **gonophores,** which often occur in clusters along a gonozooid, and a gas-filled "float," the **pneumatophore.**

The pneumatophore is a highly muscular, double-walled bladder that imparts buoyancy to the colony. Because it resides at the water surface, the pneumatophore acts as a sail and allows the organism to move to new feeding areas. The float may reach 30 cm in length and possesses a **crest** which can be raised and lowered in response to wind conditions and locomotory needs. On the floor of the float is a region of modified glandular epithelium, the **gas gland.** This gland produces the carbon dioxide and nitrogen gasses found in the float. Although lacking in *Physalia*, many related species possess a pore that regulates the amount of gas, and thus the buoyancy, of the float.

Several long dactylozooid tentacles are present that are equipped with batteries of cnidocytes. Contact with the fishing tentacles results in firing of the nematocyst and release of toxins into the wound. *Physalia* packs a potent "sting" that can be harmful or even fatal to humans. The toxins are proteinaceous so many divers carry meat tenderizers (which contain proteolytic enzymes) to neutralize the toxins.

Class Scyphozoa: *Aurelia*

Aurelia, a typical scyphozoan jellyfish, is a marine organism common along coastal waters of temperate and tropical latitudes (Figure 1.20). Although the diversity of the scyphozoa is limited (200 species), their abundance is great and they are found in all oceans, including the polar regions. Scyphozoan medusae (**scyphomedusae**) are large (typically 5–50 cm in diameter; some are as large as 2 meters) and represent the conspicuous stage of the life cycle. Their polyps are either absent or reduced to a small, inconspicuous form. The mesoglea is thick and cellular, imparting a firmer consistency than the hydromedusae. Scyphomedusae are constructed in a similar manner to that of hydromedusae, but they lack a **velum,** a shelf-like membrane used in swimming. Their body parts are arranged symmetrically around the oral-aboral axis, usually in fours or multiples of four, a form of radial symmetry termed **tetrameric.** Most scyphozoans are carnivorous, feeding on fish and small invertebrates, but *Aurelia* is a filter feeder which captures plankton.

PRESERVED SPECIMENS

General structure. Examine a specimen on demonstration. Note that *Aurelia* is **discoidal** and when spread flat, shows a circular shape disrupted at eight regular intervals by notches in the margin. Each marginal notch contains a **rhopalium,** which contains a **statocyst** and an **ocellus.** These are sensory structures responsive to a variety of stimuli. The statocyst is gravity sensitive and contains calcium sulfate granules which communicate with ciliated sensory cells to regulate balance of the organism. The ocellus is a photoreceptive pigment cup which senses differences in light intensity. Small **tentacles** circumvent the organism.

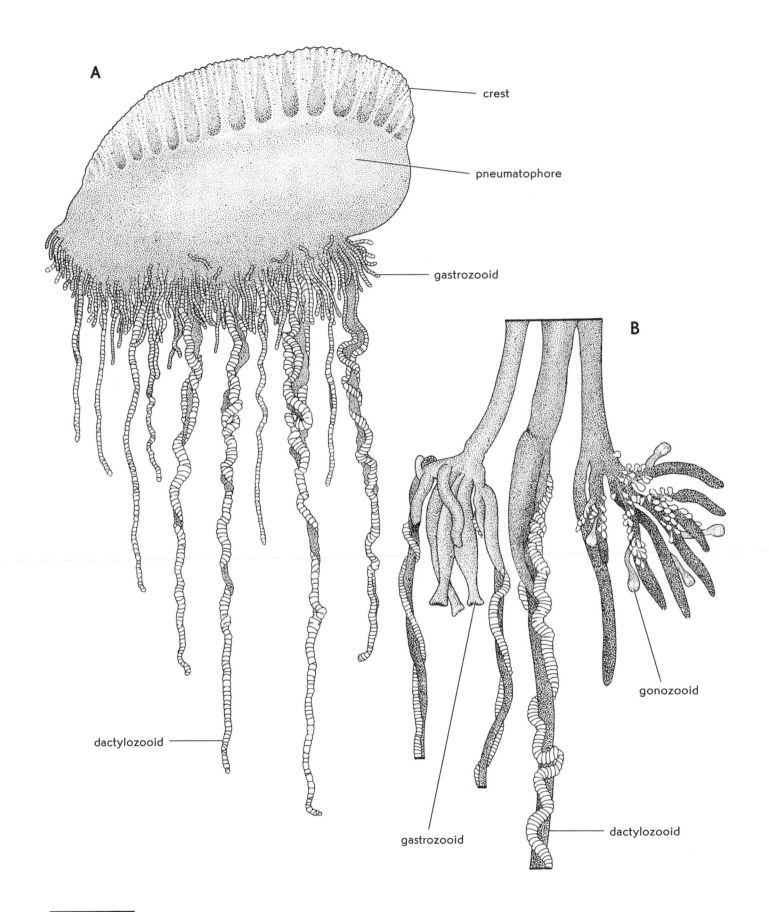

FIGURE 1.19 *Physalia,* the Portuguese man-of-war. A) entire colony; B) detail of the tentacles showing specialized zooids.

Labels in figure:
- crest
- pneumatophore
- gastrozooid
- dactylozooid
- gonozooid
- gastrozooid
- dactylozooid

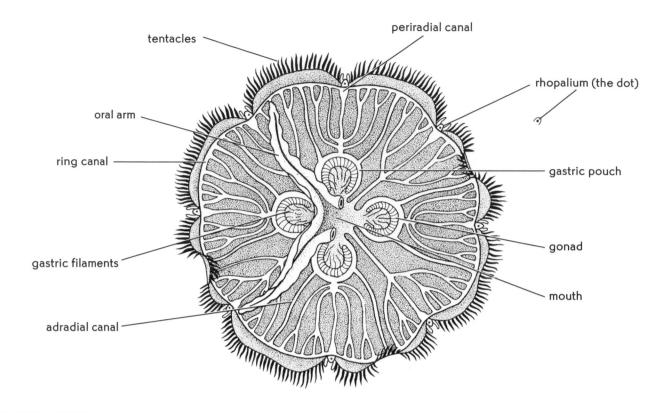

tentacles

periradial canal

oral arm

rhopalium (the dot)

ring canal

gastric pouch

gastric filaments

gonad

adradial canal

mouth

FIGURE 1.20 Structure of *Aurelia*, the "moon jelly." The oral arms have been removed from the right side.

Gastrovascular system. Observe the oral surface of the organism. Four long, tongue-like **oral arms** extend from the **mouth.** These are modifications of the **manubrium,** a sheath-like tube which gives rise to the mouth. The mouth opens into a short **gullet,** which leads to the **stomach.** Around the stomach lie four **gastric pouches** each of which contains a horseshoe-shaped **gonad.** The inner edges of the gonads constrict into numerous thin processes, the **gastric filaments.** The filaments house **nematocysts** to subdue prey items and secrete enzymes into the digestive system, the **gastrovascular cavity.** In *Aurelia* the digestive system forms a complex network of canal systems and branches. **Radial canals** run from the gastric pouches to the **ring canal** that traces the outer margin. *Aurelia* medusae trap plankton in mucus at the outer edge of the body. The plankton are removed from the margin by the oral arms and water currents created by flagella carry the particles along grooves in the arms to the stomach.

Reproduction. The adult medusa reproduces sexually and fertilization is external. Sex cells are released from the gonads into the gastrovascular cavity and are discharged through the mouth. Within the folds of the oral arms the young embryos develop into ciliated, free-swimming **planula larvae.** These eventually dissociate from the parent and settle on the sea floor. The larvae then develop into small polyps called **scyphistomae.** When favorable environmental conditions are present, the scyphistomae begin to **strobilate,** or asexually bud

off young medusae (**ephyrae**) in layers. The ephyrae transform into adult jellyfish to complete the life cycle.

Class Anthozoa: Corals and *Metridium*

Anthozoans are typically colonial polyps, called corals, which secrete a hard, calcareous external skeleton (Figure 1.21). Successive generations build upon the skeletons of former generations and many form massive underwater **coral reefs.** These reefs proliferate in tropical regions and support an incredible array of life forms. Often the reef environment is the nexus of complex marine food chains. Anthozoan polyps (the medusa form is absent in this class) are often brightly pigmented, possess a crown of tentacles surrounding the oral surface, and when disturbed, can partially withdraw into their cup-shaped shell. Anthozoans have a thick cellular mesoglea and a differentiated gastrodermis that contains cnidocytes and gonads. The gastrovascular cavity of anthozoans is divided into compartments by longitudinal mesenteries termed **septa.** These septal walls compartmentalize the body and function as a **hydrostatic skeleton** in those species which lack a calcareous shell.

Metridium

Metridium, the sea anemone (Figure 1.22), is an atypical anthozoan which is found on pilings or stones and in rocky crevices along the Atlantic and Pacific coasts.

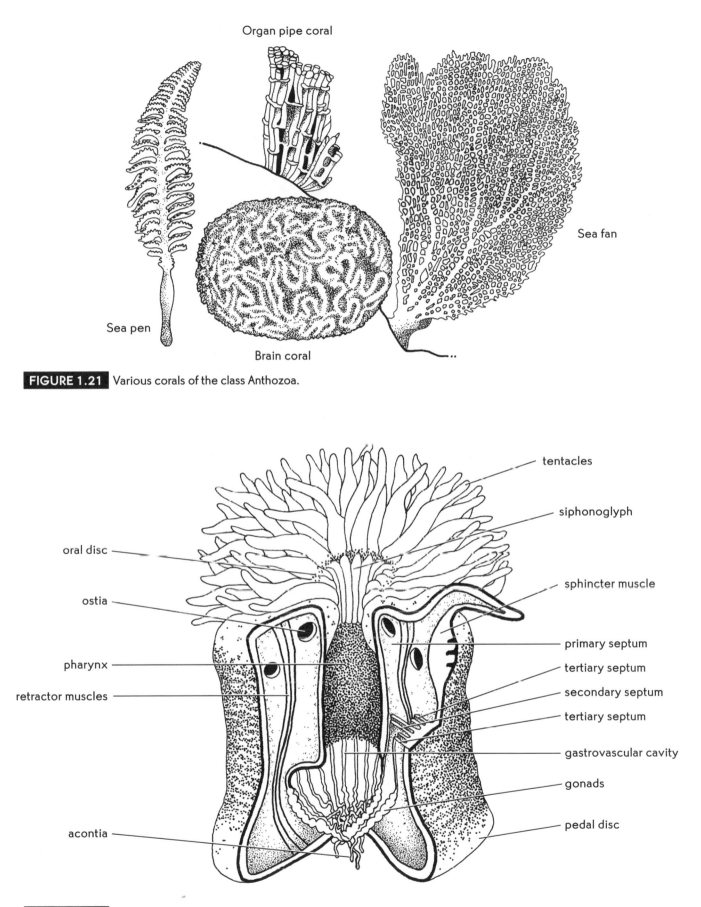

Organ pipe coral

Sea pen

Brain coral

Sea fan

FIGURE 1.21 Various corals of the class Anthozoa.

tentacles

siphonoglyph

oral disc

sphincter muscle

ostia

primary septum

pharynx

tertiary septum

secondary septum

retractor muscles

tertiary septum

gastrovascular cavity

gonads

acontia

pedal disc

FIGURE 1.22 *Metridium*, the sea anemone. The free edges of the septal walls and the acontia threads are provided with nematocysts to complete the prey paralyzation begun by the tentacles.

A few species forsake a hard substrate and burrow in soft mud or are free-swimming. *Metridium* occurs commonly in shallow water but is also known from depths of 50 to 75 meters. Sea anemones are solitary sessile animals which possess a **hydrostatic skeleton.** Unlike most other anthozoans they do not secrete an external shell and do not live in colonies, although some individuals may cluster together in prime habitats.

 **PRESERVED SPECIMENS**

External and internal structure. Observe a specimen in a dissecting pan. If a model of the organism is available use this and Figure 1.22 as a supplement. Due to the compartmentalization and pressurization of the water in the body by septal walls, *Metridium* is sturdier and more muscular than many other cnidarians. The body is cylindrical and the animal can be divided into three primary regions: the **oral disc** or free end, with numerous conical **tentacles** surrounding the **mouth;** the cylindrical **column,** forming the main body of the organism; and the **pedal disc** (aboral end). *Metridium* attaches itself to a solid object by the pedal disc. Gland cells in the pedal disc secrete a gluey substance which helps the organism adhere to the substrate. Although it is a sessile animal, the sea anemone can dislodge itself and creep slowly over the substrate on its pedal disc.

The inner surface of the mouth is lined with ridges and several ciliated grooves, the **siphonoglyphs,** are found along the sides of the mouth. The ciliated siphonoglyphs drive water into and out of the muscular pharynx, or throat region. The pharynx opens into the **gastrovascular cavity** which is large and subdivided into six **radial chambers** by **primary septa,** vertical extensions of the body wall. Note that the chambers formed by the primary septa in the gullet region communicate with each other by means of small oval holes, the **ostia.** The circulation of water into the gastrovascular cavity facilitates respiration and expels waste and foreign materials from the body. The six primary chambers are further subdivided by smaller, incomplete **secondary** and **tertiary** septa which are free at their inner edges. The free edges of the septa may bear thread-like **acontia.** The acontia threads are equipped with plenty of nematocysts to subdue struggling prey in the gastrovascular cavity. The outer covering (**epidermis**) is tough, and small pores on tiny papillae are scattered about the epidermis.

Reproduction. Sea anemones are dioecious organisms which can reproduce both asexually and sexually. Asexual reproduction can be accomplished by **fission** but more commonly occurs as **pedal laceration.** Pieces of tissue from the pedal disc are separated from the parent body as the anemone moves across the substrate. As if they were following the "footsteps" of the parent organism, small sea anemones regenerate from these tissue bits. Sexual reproduction occurs as gametes are released from the **gonads.** The gonads are located along the secondary and tertiary septa within the gastrovascular cavity. The gametes pass out of the gastrovascular cavity and fertilization occurs externally in the seawater. The **zygotes** develop into **planula larvae** which settle on the substrate and metamorphose into adults.

✓ Checklist of Suggested Demonstrations

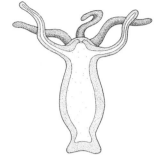

THE CNIDARIANS

_____ 1. Discharged nematocysts slide.

_____ 2. *Hydra* whole-mount slide with accompanying diagram.

_____ 3. Monoecious *Hydra* (with spermaries and ovaries) slide.

_____ 4. Budding *Hydra* whole-mount slide.

_____ 5. *Hydra* cross-section slide.

_____ 6. *Hydra* living specimens; accompanied with *Daphnia* as a food source.

_____ 7. *Physalia* preserved specimen.

_____ 8. *Aurelia* preserved specimen.

_____ 9. *Metridium* preserved specimen.

_____ 10. *Metridium* cross-section and longitudinal-section slide.

_____ 11. *Metridium* model and accompanying diagram.

_____ 12. Sea fan (*Gorgonia*) and sea plume dried specimens.

_____ 13. Sea pen preserved specimen.

_____ 14. *Meandra* (brain coral) specimen.

_____ 15. *Fungia* (mushroom coral) specimen.

_____ 16. *Renilla* (sea pansy) preserved specimen.

_____ 17. *Astrangia* (stone coral) specimen.

Cnidarian Notes and Drawings

Cnidarian Notes and Drawings

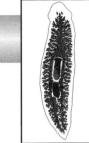

EXERCISE 2

Platyhelminthes, Pseudocoelomates, and Annelids

THE ACOELOMATE FLATWORMS

THE PLATYHELMINTHES

The platyhelminthes, or **flatworms,** are **bilateral** marine, freshwater, and moist soil organisms that lack a body cavity. These **acoelomates** are more complex than radiate animals but vastly simpler than the remaining bilateral animals. They include the free-living **turbellarians** (planarians) and the parasitic **flukes** and **tapeworms.** Of the 15,000 known species more than 85% are parasitic. Because they lack a body cavity and a complex circulatory system, their volume must remain low in comparison to their surface area. As a result they are constrained in shape and size to flattened, worm-like organisms adapted for a benthic, or parasitic existence. Flatworms range from a few millimeters in length to 15–20 meters in some tapeworms. Acoelomates can attain such great length only because their dorsoventral flattening reduces the distance that gases, wastes, and nutrients must flow.

They are the least advanced animal form which possesses **organs,** and many also possess complicated **organ systems.** The reproductive system is well developed and much of the energy produced by the organism is channeled into reproduction. However, they still lack true circulatory and respiratory organs, and the gastrovascular cavity (when present) lacks an anus. The digestive tract is incomplete with a single two-way opening, the mouth. The formation of these organs is due to the presence of a true **mesoderm.** Indeed, acoelomates are **triploblastic** (three embryonic germ layers) and most of their mesoderm is derived from endoderm. Because the interior is packed with **mesenchyme,** diffusion is no longer an efficient method for waste elimination; nitrogenous wastes are excreted by a specialized organ, the flame cell or **protonephridium.**

Flatworms are **cephalized** with a concentration of nervous tissue and sensory organs in a true head. Cephalized acoelomates move headfirst through the environment to maximize the efficiency of cephalization. It is thought that flatworms are distantly related to the protostome molluscs, annelids, and arthropods but because most species are adapted for parasitism these modifications have obscured their evolutionary relationships.

The group is distinguished by several unique features: biflagellate sperm cells, an atypical 9 + 1 arrangement of microtubules in sperm flagella, and **vitellaria,** yolk producing structures separate from the ovary.

Class Turbellaria: *Dugesia*

Turbellarians are free-living predators and scavengers with a typical flat, leaf-like shape (Figure 2.1). They are

---

SYSTEMATIC OVERVIEW

PHYLUM PLATYHELMINTHES

Flatworms; bilateral acoelomates; 15,000 species (13,000 of which are parasitic).

Class Turbellaria

Most are free-living, including planarians; a few are commensals on marine invertebrates (*Dugesia*).

Class Trematoda

Digenetic flukes; they are all parasitic species, with both asexual and sexual reproduction (*Clonorchis, Fasciola, Schistosoma*). Monogenetic flukes; all species are parasitic with only sexual reproduction (*Gyrodactylis, Polystoma*).

Class Cestoda

Tapeworms; all species are parasitic and gutless; hydatid cyst larvae reproduce asexually (*Taenia, Dipylidium, Echinococcus*).

PHYLUM GNATHOSTOMULIDA

Common in anaerobic beach sands; 80 species (*Austrognathia, Gnathostomula*).

PHYLUM NEMERTEA

Ribbon worms; 800 species (*Amphiporis, Prostoma, Geonemertes*).

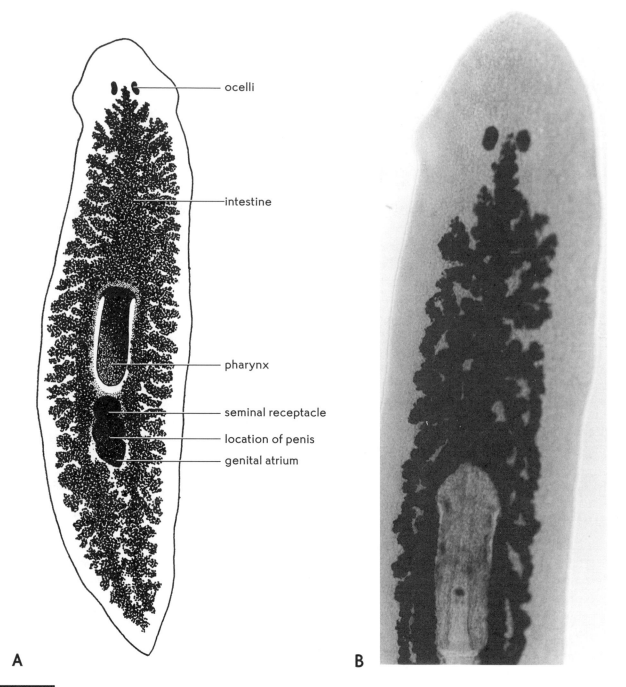

A

B

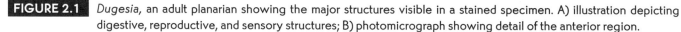

FIGURE 2.1 *Dugesia,* an adult planarian showing the major structures visible in a stained specimen. A) illustration depicting digestive, reproductive, and sensory structures; B) photomicrograph showing detail of the anterior region.

The figure labels read:
- ocelli
- intestine
- pharynx
- seminal receptacle
- location of penis
- genital atrium

common in marine and freshwater environments, but a few terrestrial species exist, making their living under wet humus among the decaying plant and animal matter. These terrestrial species are mainly tropical and subtropical and are most conspicuous after heavy rains. *Dugesia,* the planarian, is a freshwater turbellarian that lives on the underside of rocks, leaves, and sticks in lakes and ponds.

LIVING SPECIMENS

Locomotion. Place a living specimen of a freshwater turbellarian, *Dugesia,* in a watch glass along with enough pond or culture water to allow movement. Note the gliding movement produced by the action of **cilia** on the ventral surface which are coordinated with rhythmic muscular contractions of the body. Turn the animal over to its ventral surface and note its righting reaction. Increase the light intensity on the specimen and then narrow it to a reduced field and observe the reaction.

Behavior. Flatworms have a simple, uncomplicated nervous system which facilitates experimental analysis regarding orientation, sensory systems, and learning behavior. Because the nervous tissue is concentrated in the head, try the following experiments both with the head intact and with the head removed and observe the differences. First, establish the response to touch. Flatworms are negatively **thigmotactic** (touch sensitive) on the dorsal surface and are positively thigmotactic ventrally. If both dorsal and ventral surfaces are contacted planarians will not come to rest. You can observe this phenomenon by placing a piece of tin foil on the dorsal surface of a planarian. Planarians will perform righting reactions as noted earlier. Turn a worm over and observe the response. Time the righting reaction of planarians with and without the head intact. Drop planarians into a beaker with water several inches deep and note which surface contacts the container first.

Planarians are also **rheotactic** (react to water currents). Using a glass pipette, determine the reactions of planarians to a stream of water from a pipette directed at the anterior and posterior ends, as well as laterally.

Planarians are negatively **phototactic.** Allow a planarian to glide along in subdued light and then increase the light intensity on the anterior end only of the animal. Note the direction in which the animal turns. Next, allow the animal to crawl in subdued light and illuminate the sides of the organism. Repeat these experiments with animals that have one ocellus removed and then with both ocelli removed. The ability of planarians to discriminate between light and dark areas can be demonstrated by placing them in a petri dish over a piece of paper that has an arrangement of simple patterns. Checkerboard or concentric rings of dark and light areas show the best response. Observe the movements of animals against different backgrounds, both with and without the ocelli intact.

Test the ability of planarians to detect food items by placing some liver extract or bouillon near a specimen. Try this experiment with auricles intact, with one auricle removed, and with both auricles removed.

Regeneration. Regenerative powers are well developed in many turbellarians but are absent in adult trematodes and cestodes (except for the scolex and neck region of tapeworms). *Dugesia* exhibits remarkable regenerative capabilities and is a prime example in the study of morphogenesis and polarity. Dip a razor blade in acetone to sterilize it. Prepare a petri dish for operated specimens by labeling it and adding culture water. A watch glass can be used for discarded pieces. Operate while observing under a binocular dissecting microscope. With a camel hair brush, place a starved specimen

of *Dugesia* in a drop of water on a clean microscope slide. Place the slide on top of an ice cube to minimize the movements of the worm. Make perpendicular or longitudinal incisions with the razor blade, not oblique incisions. To ensure regeneration rather than healing of the incision, the cut surface should be clean and thorough. After the operation transfer the cut pieces to the labeled petri dish. Do not feed them, and store them in a cool dark place. You, or the lab instructor, may need to change the culture water in a few days. Regeneration will be completed in eight to fourteen days.

Many possible cuts can be made which result in different and interesting patterns. Frequently performed operations include: (1) Two heads; cut off the head below the ocelli, discard the head, and slit the body longitudinally to the pharynx. The slit may have to be reopened in the next twelve to twenty-four hours to ensure the formation of two heads. (2) Double head; cut a piece from the center of the body, including the posterior half of the pharynx, discard the head and tail and keep the central portion. Each side should form a new head. (3) Two tails; slit an organism from the tail down the center as far as the eyes. Twenty-four hours later, reopen the slit and cut off the head. A new head and a new tail should form from the two ends.

PREPARED SLIDES

External anatomy. Examine the specimen with a dissecting microscope and note the overall appearance. It is bilaterally symmetrical and flattened dorsoventrally. The pigmentation is irregular and clusters about the body. The **head** is distinct from the body and contains sensory organs, the **ocelli** and the **auricles.** The ocelli are pigmented eyespots that sense light intensity but do not produce visual images. The auricles are chemoreceptors which can also respond to touch. Internally, the head region houses a concentration of nerve ganglia which process sensory information.

Turn the animal over on its ventral surface. In the middle of the animal the **mouth** and the muscular **pharynx** may be visible. Small, somewhat transparent individuals may show pockets of the **gastrovascular cavity** which run anteriorly and posteriorly from the pharynx. Planarians are carnivorous animals that prey on protozoans, rotifers, and small crustaceans. Food items are ingested in the muscular pharynx where proteolytic enzymes penetrate the prey and liquify the contents. The contents of the prey can be sucked into the gastrovascular cavity for assimilation. Digestion is both **extracellular** and **intracellular** as enzymes are dumped by **gland cells** into the gastrovascular cavity and subsequently small bits of food are engulfed by **phagocytic cells** in the lining of the **gastrodermis.**

Internal anatomy. Examine a prepared slide of a series of cross-sections (anterior, pharyngeal, and posterior) of *Dugesia*. In all three notice the one-layered, cellular **epidermis**; in other species the epidermis may be syncytial. The ventral surface is ciliated to aid in locomotion. Observe the anterior section and look for the main branches of the **gastrovascular cavity**; they will appear as pockets interspersed within the tightly packed **parenchyma**. In the pharyngeal region look for the gastrovascular cavity and the more prominent muscular walls of the **pharynx**. Also identify the **circular, longitudinal,** and **oblique muscles** which cross the parenchyma. Two white, circular **ventral nerve cords** are also present.

Class Trematoda: *Clonorchis, Fasciola,* and *Schistosoma*

 **PREPARED SLIDES**

Clonorchis

Trematodes, or **flukes**, are all parasites. Most flukes are **digenetic** because their life cycle includes both sexual and asexual reproductive stages. Larval flukes undergo asexual reproduction and infect their **intermediate** host, usually a mollusc. Adults produce gametes and parasitize a variety of vertebrates (the **definitive** host). **Monogenetic** flukes (whom some authors consider a separate class) resemble digenetic flukes but reproduce only sexually. Monogenetic flukes infest the body surface of vertebrates, most notably the gills of fish, or even the eyes of hippopotamuses. *Clonorchis* (= *Opisthorchis*) is a common digenetic fluke which inhabits the bile ducts of human livers. Here it feeds on liver cells and blood and heavy infections can impair the functioning of the liver. Most are unaffected by drugs, but infection rarely results in death of the hosts.

External and internal anatomy. Examine a whole mount of the adult liver fluke *Clonorchis* (Figure 2.2). Under a dissecting microscope or low power of a compound microscope, observe the anterior end and the **oral sucker** that surrounds the **mouth.** Extending posteriorly from the mouth are two branches of the gastrovascular cavity, the **intestines.**

The **reproductive system** comprises a tubular network that dominates the appearance of the body. Each individual is an hermaphrodite that possesses both male and female sex organs. The male reproductive organs are the **testes, vas deferens, seminal vesicle,** and a **copulatory organ.** The female reproductive system consists of an **ovary, vitellaria,** a **uterus,** a **seminal receptacle,** and a **gonopore.** Fertilization is internal and sperm is

stored in the female seminal receptacle and released when required.

Life cycle. Human infections occur primarily from ingesting raw or poorly cooked fish. The eggs of the animal pass out of the host with the feces and fecal contamination of water sources spreads the disease. The eggs are ingested by some species of snail (**first**

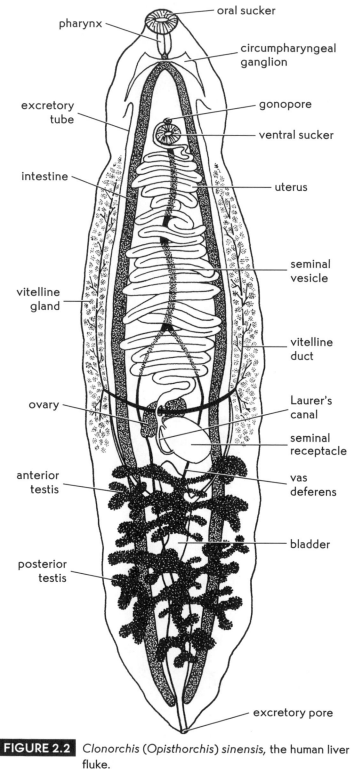

FIGURE 2.2 *Clonorchis (Opisthorchis) sinensis,* the human liver fluke.

intermediate host) where the first larval stage (the **miracidium**) hatches in the digestive tract. The miracidium passes through several larval stages (the **sporocyst, redia,** and the **cercaria**) in the tissue of the snail. These larval stages reproduce asexually and form thousands of offspring. The cercaria stage escapes the host and migrates in the water until it contacts the **second intermediate host,** a fish. The cercaria burrow into the skin of the fish and encyst, forming **metacercaria** larvae. If the metacercaria are ingested by man or another large vertebrate (**definitive**) host, they migrate to the liver and the life cycle is complete.

Fasciola

Observe a preserved specimen or stained slide of *Fasciola hepatica,* the sheep liver fluke (Figure 2.3). As an adult, it lives in the bile ducts of many herbivores, particularly sheep, pigs, and cows and degrades the adjacent liver tissue causing "liver rot." Human infection of this fluke is rare due to infrequency of exposure rather than physiological barriers.

External and internal anatomy. *Fasciola* is much larger than *Clonorchis* and generally similar in structure,

although the reproductive system of *Fasciola* is slightly more complex. Examine Figure 2.3 and compare the structures with those of *Clonorchis* (Figure 2.2). Locate the attachment organs, the **anterior sucker** and the neighboring **ventral sucker.** The **genital pore** lies in between the two suckers.

The digestive system includes the anterior **mouth,** the **pharynx,** the **esophagus,** and the branches of the intestine. The intestinal network arises as two main branches which ramify into smaller lateral branches with many diverticula.

The reproductive organs include the male **testes, vas deferens, seminal vesicles, penis,** and the **ejaculatory duct.** The **ovary, yolk glands (vitellaria), yolk ducts, oviducts,** and the **uterus** comprise the female organs.

Schistosoma

Although most trematodes are monoecious, there are dioecious flukes called schistosomes which live in the hepatic portal system of birds and mammals. These sexually dimorphic blood parasites cause severe health problems for their hosts. Three species of *Schistosoma* cause disease in humans, residing either in the veins of

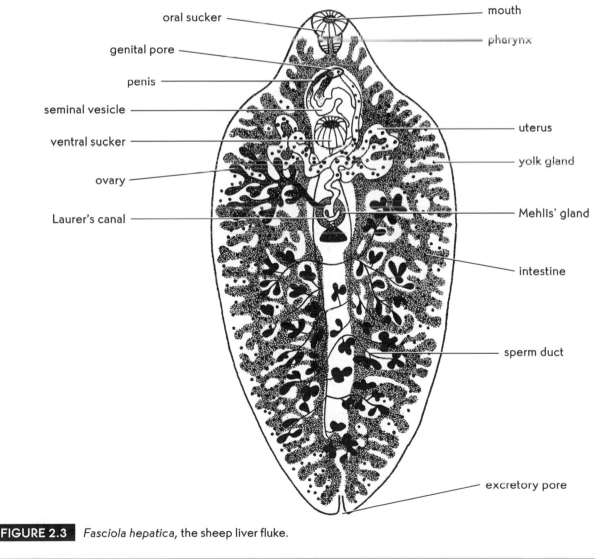

FIGURE 2.3 *Fasciola hepatica,* the sheep liver fluke.

the intestine or the bladder. Both the male and female worms live simultaneously in the host's blood vessels where their eggs clog and occlude capillaries, damage internal organs, and cause internal allergic reactions. Drugs can kill the adult worms but the eggs survive chemical treatment. Examine demonstration slides of the human blood fluke *Schistosoma* showing the male and female worms in **copula.** If available, observe one or several of the larval stages.

Class Cestoda: *Taenia* and *Echinococcus*

Taenia

The cestodes comprise 4,000 species of endoparasitic **tapeworms.** Like most flatworms, the elongate tapeworms have a ribbon-like shape; however, they differ from most platyhelminthes in a number of features, most notable is the complete lack of a digestive tract. Tapeworms can reach lengths of up to several meters; the smallest species are slightly less than 10 mm. Adult

tapeworms are common infectious agents of vertebrates but the adults rarely cause significant harm to their hosts. The larvae on the other hand, migrate through their hosts causing immense tissue damage. *Taenia* is a common genus of cat, dog, and pig tapeworms that inhabit the small intestine of their host (Figure 2.4). Larval stages of *Taenia* are found in the liver and mesenteries of rabbits.

General structure and reproductive system. Examine the tapeworm *Taenia* (Figure 2.4) and note the linear series of repeating segments, or **proglottides,** which become increasingly larger towards the posterior. The anterior end sports the attachment organ, the **scolex.** The microscopic scolex is equipped with piercing **hooks** and strong **suckers** which dig into the intestinal wall of the host. Posterior to the scolex is the **neck,** the site of newly

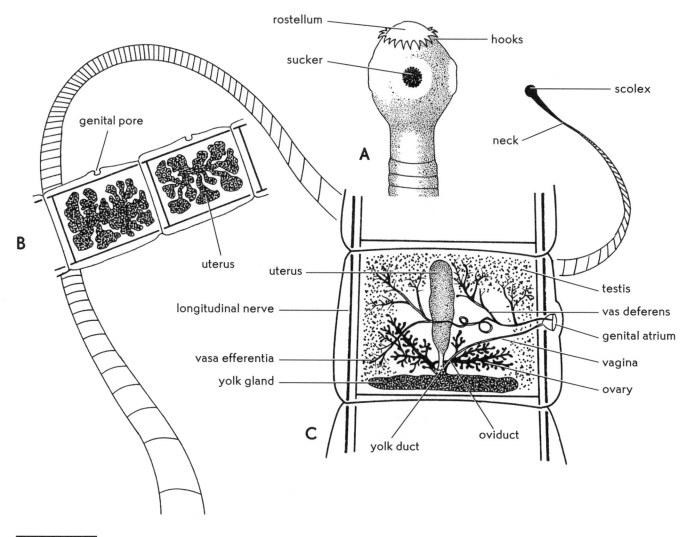

FIGURE 2.4 *Taenia solium,* the pork tapeworm. A) scolex showing hooks and suckers; B) gravid proglottides showing uteri packed with eggs; C) structure of a mature proglottid.

formed, or **immature**, proglottides. Proglottides form asexually by budding in the neck region; thus, the youngest proglottides are directly behind the scolex and the oldest proglottides extend posteriorly. When **mature**, each proglottid houses both male and female reproductive organs. A mature proglottid contains the female **uterus, yolk glands, oviduct, vagina,** and **seminal receptacle.** The male structures consists of the **testes, vas efferens, vas deferens,** and the **cirrus,** a copulatory organ. A common **genital pore** services both sets of reproductive organs. Self-fertilization is common among tapeworms. **Gravid** proglottides, found at the posterior end, are little more than a uterus completely distended with eggs. They become brittle with age and are shed with the feces; on drying outside the body, the **eggs** are liberated.

Study a prepared slide of a mature proglottid of *Taenia* and identify the structures of the male and female reproductive system. A quick glance at a prepared slide of gravid proglottides of *Taenia* will show the uterus packed with eggs.

Note that the proglottid structure is dominated by the reproductive tract. The digestive tract is nonexistent, so there's not even a mouth. The tapeworm obtains its nourishment by simple **diffusion** of nutrients across its outer surface. This accomplishment is facilitated by the adult habitat, the digestive tract of its host. What better place to encounter dissolved nutrients?

Due to the nature of proglottid structure, tapeworms appear to be segmented. However, most zoologists regard a tapeworm as a "colony" or chain of individuals (proglottides) because each "segment" is capable of reproducing itself. Also, due to peculiarities in development, proglottides are not officially regarded as true segments.

Larval stages. When *Taenia* eggs are eaten by an **intermediate host,** such as a flea, a six-hooked embryo (**oncosphere**) emerges, penetrates the gut of the host, and encysts in the tissues as a **bladder worm** or **cysticercus.** The cysticercus remains in this stage until the intermediate host is eaten by a **final host,** usually a dog or cat. In the digestive tract of the final host the scolex emerges and a young tapeworm develops.

Echinococcus

The adult tapeworm *Echinococcus* lives in the gut of canine carnivores (Figure 2.5). Ingestion of eggs by a suitable mammal results in the development of a cysticercus with multiple bladders (**hydatid cyst**) in the liver, lungs, or the brain, where their cyst formation causes considerable damage. The cysts are resistant to drugs and must be removed surgically. Humans are potential intermediate hosts for *Echinococcus,* where it infects the lungs and causes large cysts to form. Examine a slide of both a tapeworm cysticercus and a hydatid cyst, noting the similarity of the inverted scolices.

cysts

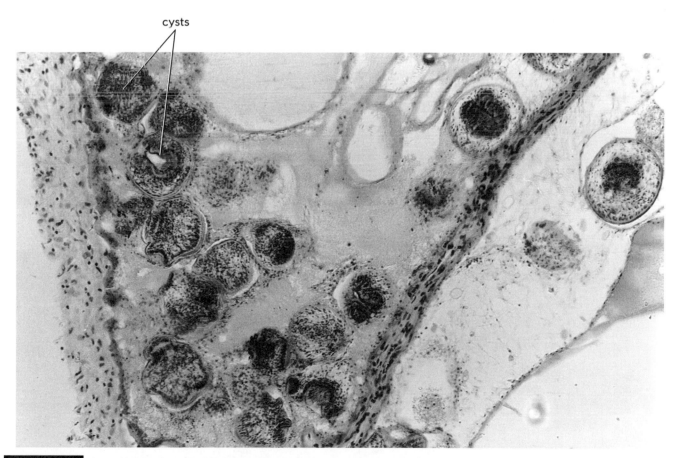

FIGURE 2.5 Larval hydatid cysts of *Echinococcus,* a tapeworm which can infect a variety of carnivorous mammals.

✓ Checklist of Suggested Demonstrations

THE PLATYHELMINTHES

_____ 1. *Dugesia* whole-mount slide.

_____ 2. *Dugesia* anterior, pharyngeal, and posterior cross-section slide.

_____ 3. *Dugesia* living specimens in culture.

_____ 4. *Bipalium* (land planarian) preserved specimen.

_____ 5. *Clonorchis* whole-mount slide and accompanying diagram.

_____ 6. *Clonorchis* life cycle diagram.

_____ 7. *Fasciola* whole-mount slide and accompanying diagram.

_____ 8. *Fasciola* life cycle diagram.

_____ 9. *Schistosoma* miracidium larva whole-mount slide.

_____ 10. *Taenia* preseved specimens.

_____ 11. *Taenia* scolex slide.

_____ 12. *Taenia* proglottides slide with accompanying diagram.

_____ 13. *Echinococcus* cyst slide.

Platyhelminth Notes and Drawings

Platyhelminth Notes and Drawings

THE PSEUDOCOELOMATE PHYLA

SYSTEMATIC OVERVIEW

PHYLUM NEMATODA

Roundworms; free-living or parasitic; generally internal parasites known as hookworms, heartworms and filarial worms; 90,000 species (*Ascaris*, *Ancylostoma*, *Dirofilaria*, *Necator*, *Turbatrix*, *Wuchereria*).

PHYLUM NEMATOMORPHA

Horsehair worms; adults are free living and larvae are parasitic in arthropods; 240 species (*Gordius*).

PHYLUM ACANTHOCEPHALA

Spiny-headed worms; adults have a spiny invertible proboscis and no gut; all are parasitic, mostly in the gut of fishes; larvae and juvenile infect arthropods; occasionally infect people; 700 species (*Macracanthorhynchus*).

PHYLUM ROTIFERA

Rotifers; a ciliated corona surrounding the mouth and a muscular pharynx (mastax) with jaw-like parts are diagnostic features; the cuticle is intracellular; parthenogenesis is common; freshwater and marine; 1,800 species (*Philodina*).

PHYLUM GASTROTRICHA

Gastrotrichs; body is covered with spines and has two forked adhesive tubes at the posterior end; pseudocoel reduced to tiny spaces or absent; parthenogenesis common; occur in freshwater and marine environments; 450 species (*Lepidodermella*).

PHYLUM KINORHYNCHA

Kinorhynchs; all are marine; evidence of segmentation in the spiny cuticle, muscles, and nervous system; the head is retractable; 100 species (*Echinoderes*).

PHYLUM LORICIFERA

Loriciferans; marine sand dwellers; body mostly surrounded by an armored lorica; few species are as yet identified (*Nanaloricus*, *Pliciloricus*).

PHYLUM PRIAPULIDA

Priapulids; bottom-dwelling marine worms with a barrel-like anterior proboscis adapted for carnivory and deposit feeding; 13 species (*Priapulus*, *Tubiluchus*).

PHYLUM ENTOPROCTA

Entoprocts; solitary or colonial; individuals are sessile and stalked; mouth and anus are situated within a crown of tentacles; many are colonial; there is one freshwater genus; 100 species (*Barentsia*, *Loxosoma*, *Pedicellina*, *Urnatella*).

The pseudocoelomates are a heterogenous assemblage of organisms which share a few common characteristics. Formerly united as a single phylum (the **aschelminthes**), these organisms probably are not very closely related. They are linked by the presence of the prominent **pseudocoel,** a body cavity that is not completely lined by mesodermal tissue. Among others, the group includes the notorious **nematodes** or roundworms, as well as the **rotifers, gastrotrichs,** and the **nematomorphs.** These phyla are all aquatic, including those parasitic species which live in the body fluids of their host, and the soil species which live in water droplets between soil particles. All have a complete digestive tract and an outer covering of the body, a nonliving, protective **cuticle.** Pseudocoelomates lack specialized circulatory and respiratory organs, and these functions are performed by diffusion through the fluid-filled pseudocoel. The pseudocoel also houses the well-developed reproductive and excretory organs.

In some pseudocoelomates, particularly the rotifers, cell division ceases in the somatic organs sometime during development. This results in **eutely,** a rare condition where adult individuals have a fixed number of body cells. Growth continues as an increase in cell size rather than number. Eutelic species have a set number of adult body cells, but perfect eutely is rarely achieved because certain somatic cells occasionally continue to divide during growth.

Phylum Nematoda: *Ascaris*

Nematodes, with 90,000 extremely abundant species, are some of the most ecologically important animals on the planet. Due to their incredibly small size, most people are not familiar with these common worms, despite frequent contact with them. A handful of fertile soil may contain several thousand microscopic nematodes. A single decaying apple was known to harbor over 100,000 individuals. They are known to exist in a great variety of habitats from Arctic ponds to desert soils, from deep sea muds to mountain springs. Their ability to parasitize is unequaled, for they are known to inhabit every vertebrate species. Humans are plagued by 15 species of pathogenic nematodes, but an equal number may opportunistically enter parasitic relationships with man.

Despite their diversity, nematodes are limited in shape and form. Nearly all species are elongate and cylindrical with tapering ends. The phylum's success is largely a measure of the adaptability of this architecture for parasitism. They range in size from less than 1 mm in length to over several meters. *Ascaris* is a nematode parasite of the intestines of man, pigs, and horses. Reaching a length of up to 50 cm, it is considerably larger than most roundworms, making it appropriate for dissection and the depiction of nematode body structure. Common infections occur in third world countries and places of poor sanitation. The *Ascaris* infection rate approaches 15% even in some areas of the United States.

PREPARED SLIDES

External anatomy. Examine the external appearance of a preserved *Ascaris* (Figure 2.6). Note that the body is covered by a thin superficial layer, the transparent, acellular **cuticle.** The cuticle is secreted by the underlying syncytial layer of cells, the **hypodermis.** The body is unsegmented, although sometimes the cuticle may be ridged. Ascarids are dioecious and sexually dimorphic. The male is smaller (up to 30 cm) and has a hooked posterior end with small **copulatory spicules** projecting from it. The female is larger (up to 50 cm) and thicker. Be sure not to confuse the undulating form of the female with the sharp, distinct hook of the male posterior. Note that the dorsal and ventral midlines are marked by narrow **longitudinal lines,** and broader marks are found laterally (**lateral lines**). The **mouth** is a small opening located at the anterior end and is surrounded by three (or six) **lips.** The **excretory pore,** located just below the mouth region, drains the inner excretory ducts. Female worms have a ventral **genital pore** located about a third of the way down the body from the mouth. The **anus** appears as a transverse slit on the ventral surface near the posterior end.

Internal anatomy. The body wall is thin and flexible, making dissection a simple but meticulous event. Insert a straight pin at each end of the animal to secure the specimen to the dissection pan. With a razor blade, sharp scalpel, or even the point of a pin, make a single longitudinal cut from end to end. Be careful not to make this cut too deep or it will disturb the internal organs. Pull back the body wall and insert pins obliquely to hold the wall away from the internal organs. Examine the arrangement of the organs under the dissecting microscope. The organs lie free in the pseudocoel and are not bound by mesenteries. Note the tube-within-a-tube structure. The internal anatomy is simple and the digestive, reproductive, and excretory systems are visible as long tubes. The **intestine** is thin and ribbon-like and runs nearly the entire length of the body. The **excretory ducts** are two small thin tubes which diverge from a main branch near the anterior end. These ducts lie within the lateral lines and may require careful inspection for identification. The remaining tubular organs comprise the reproductive system; these will be studied in detail in the next section.

Reproductive system. Most nematodes are **dioecious,** with separate individuals for the sexes. Both the male and the female reproductive systems consist of long, coiled tubes which increase in diameter from the gonad to the external opening. Male systems may be paired (as in the female) or unpaired. This, unfortunately, has caused much confusion among zoology lab students and instructors. Whether single or paired, the reproductive system consists of **testes, vas deferens,** and **seminal vesicles** which open into the posterior end of the intestine. The male spicules projecting from the posterior end facilitate copulation by holding the female vagina open while amoeboid sperm are pumped into the reproductive tract. The female tract possesses paired **ovaries, oviducts,** and **uteri** as well as a single **vagina.** The vagina opens to the genital pore. Near the junction of the oviduct and uterus is the **seminal receptacle,** where fertilization takes place. Some sections of the reproductive tract, for example, the uterus and vagina, are capable of **peristaltic action,** waves of muscular contraction. The amoeboid sperm and peristaltic extrusion of eggs are both functional adaptations designed to overcome the high turgor pressure in the pseudocoel.

Life cycle. Because they lack larval stages, nematodes have fairly simple life cycles. Following copulation, fertilization takes place in the seminal receptacle, and the females lay eggs containing developing juveniles. The eggs continue to develop in the soil where they may be ingested by a new host organism. The eggs are extremely hardy and remain viable for years under the most adverse of conditions. The eggs are adapted to combat temperature extremes and desiccation and they are also known to survive immersion in preservatives (such as formaldehyde) for years. Make sure you wash your hands after dissecting an *Ascaris* and don't put your fingers in your mouth while doing it! You would not be the first student to pick up an *Ascaris* infection after contact with it in lab! The eggs hatch in the small intestine and the juveniles are essentially miniature adults, albeit sexually immature. The juveniles migrate through various tissues of the host and pass through several transitory stages before becoming adults and reentering the small intestine. Most nematodes infect only one host species, and often only the juveniles or adults are parasitic.

Phylum Nematomorpha: horsehair worms

A small group of freshwater pseudocoelomates, the nematomorphs are slender, elongate, cylindrical worms. Most of the 240 species range in length between 10 and 100 cm, but all are only a few millimeters in diameter. Only a single genus is marine, and no strictly terrestrial forms exist. The juveniles are parasitic on arthropods or leeches, but the adults are free-living. Neither the juveniles nor the adults have a digestive tract; the larvae absorb the body fluids of the host across their body surface, and the adults do not feed. The juveniles develop underneath the cuticle of their host, and those

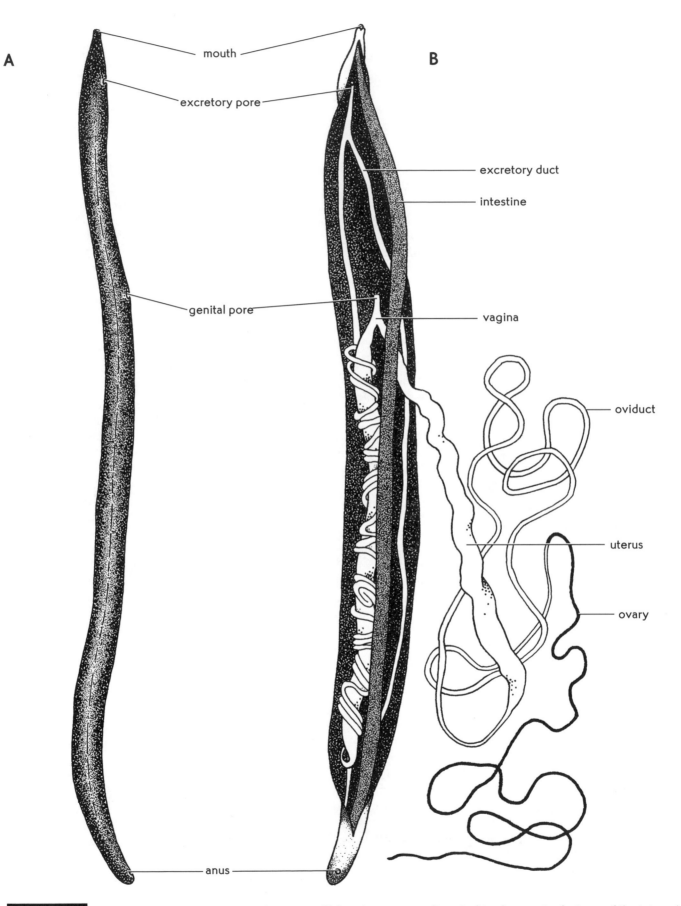

FIGURE 2.6 *Ascaris lumbricoides.* A) external anatomy; B) female specimen dissected to show major features of the internal anatomy.

that develop in terrestrial arthropods wait until the host is near water to emerge. As with the nematodes, horsehair worms lack circular muscles in their body wall, so their movements are slow and unproductive. The sexes are separate and males are smaller and more active than females. They have been known to appear in tap water because they are thin enough to pass through water treatment facilities. Fortunately, as adults, nematomorphs are completely harmless.

Phylum Rotifera: *Philodina*

Rotifers are small pseudocoelomates which superficially resemble some protozoans in many aspects of their biology. Most species are microscopic; the largest species approach several millimeters. They are important organisms in aquatic ecosystems and constitute a significant food source for many species of fishes and filter-feeding animals. Over 90% of the phylum's 1,800

species inhabit freshwater lakes, ponds, and slow-moving streams, however, rotifer species are also known from marine and even moist terrestrial situations. *Philodina* is a North American rotifer which is readily available and possesses most of the characteristics typical of the phylum (Figure 2.7).

LIVING SPECIMENS

External and internal anatomy. With a pipette, remove several drops of culture water and place in a small watch glass under a dissecting microscope or under low power of a compound microscope. Scan the medium for several small, erratically moving translucent organisms. They are most easily found attached to debris at the bottom of the dish. When a suitable specimen is found, increase the magnification and isolate the specimen. The body is not worm-like as is typical of many

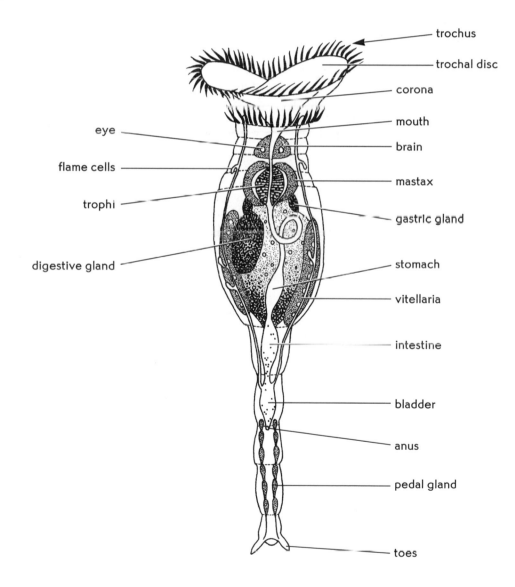

FIGURE 2.7 *Philodina roseola*. General structure of a female rotifer.

pseudocoelomates. The organism appears to be divided into segments, but in reality these are mere constrictions in the outer covering, the **cuticle**. The sectioning of the cuticle allows for telescoping which is useful in movement and feeding. This cuticular segmentation is not reflected by the internal organs and is therefore referred to as **pseudosegmentation**. The body is divided into three main regions, the **head**, the **trunk**, and the **foot**. The anterior end houses the unique **corona**, a circular crown of cilia useful in feeding and locomotion. The corona rotates, a feature evident in the name of the phylum, almost constantly, and in *Philodina* it consists of two large **trochal discs**. **Retractor muscles** operate the corona. The head region also bears a sensory organ, the **rostrum**. The **mouth** is located at the base of the corona and it receives food particles swept down by ciliary action. The muscular **pharynx** encloses a grinding apparatus, the **mastax**. The mastax is quite conspicuous and during feeding masticates back and forth as it breaks down plankton and detritus. The action of the mastax is very rhythmic and this led Leeuwenhoek, the Dutch lens maker and discoverer of protozoans, to mistakenly believe that this was the organism's heart.

The trunk region bears the internal organs associated with digestion and reproduction. The **esophagus** is surrounded by two large **salivary glands** as it connects with the thick-walled **stomach**. Digested food is absorbed by the gastrodermal lining of the stomach before the remains pass into the short **intestine**. Body wastes are held in the **cloaca** and expelled through the dorsal **anus**.

The tapered foot possesses movable **toes** and **pedal glands** which secrete an adhesive substance used in attachment to substrates. Rotifers can temporarily attach themselves to an object, elongate and contract, then dissociate from the substrate and actively swim.

Feeding. Rotifers will actively feed on cultures of the single-celled algae, *Chlamydomonas*. If this is not available, try staining some dried yeast (with Congo Red or Safranin O) and suspending it in the culture water near a few rotifers. Note the feeding response and the movements involved.

Reproduction. The reproductive strategies of rotifers is one of the most interesting and unique styles of propagation. Most natural populations consist solely of females and reproduction is accomplished through **parthenogenesis** (development from an unfertilized egg). Males either do not occur at all or exist only for short periods of time during the growing season. The males that exist are small and weak with a degenerate digestive tract. The can swim quickly, however, and they possess well-developed sensory structures. There are two general types of female rotifers; one kind always reproduces parthenogeneically (**amictic**), producing eggs which all develop into females, whereas the other kind may reproduce bisexually. The bisexual females produce males intermittently, as well as resting eggs. The cycle commences with parthenogenetic reproduction and is terminated by a short period of sexual reproduction, the timing of which depends to a large extent on environmental conditions. The female reproductive system in *Philodina* consists of a single **ovary** and a yolk-producing **vitellarium**. The **eggs** pass from the **oviduct** into the **cloaca** and then are released to the environment.

Phylum Gastrotricha: *Lepidodermella*

The gastrotrichs superficially resemble rotifers. They are of similar size (less than 1 mm) and shape, but they lack the distinctive corona. The body is flattened dorsoventrally and **cilia** cover the ventral surface (gastrotrich refers to "stomach-hair"). The cuticle is extracellular and highly modified to form spines, bristles, and scales in many species. The pseudocoel is greatly reduced or even completely absent. Like the rotifers, they have a forked posterior end which bears adhesive **pedal glands** (Figure 2.8). Gastrotrichs comprise 450 known species which inhabit marine and freshwater environments. Living on submerged plants and among sand grains, gastrotrichs feed on organic debris, bacteria, protozoa, and algae. Conversely, gastrotrichs are an important food source for amoebas, hydras, nematode and annelid worms, and even small arthropods. Cilia surround the mouth and are used in the sorting of food particles. Food is pumped into the body by action of the muscular **pharynx**.

Gastrotrichs are monoecious, but marine species copulate to ensure genetic diversity. Freshwater species reproduce **parthenogenetically** or they may cross fertilize. Examine slides or demonstrations of living gastrotrichs, and if available, compare the form of an acanthocephalan or other groups of pseudocoelomates listed previously but not discussed (Figure 2.9).

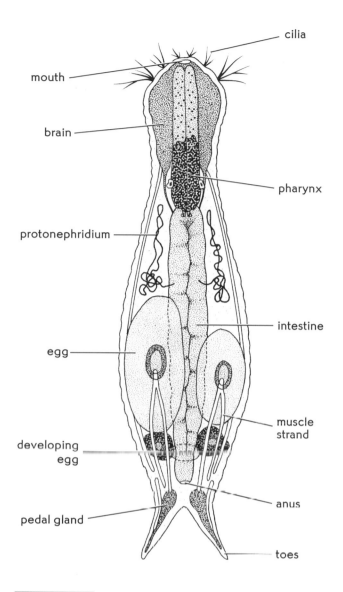

FIGURE 2.8 Internal features of *Lepidodermella,* a freshwater gastrotrich.

Labels (Figure 2.8): cilia, mouth, brain, pharynx, protonephridium, intestine, egg, muscle strand, developing egg, anus, pedal gland, toes

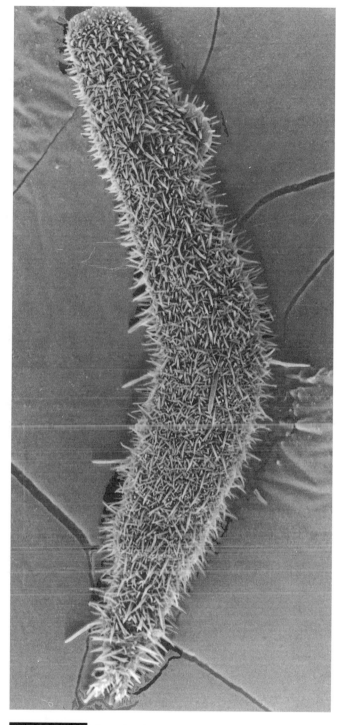

FIGURE 2.9 *Acanthodasys* sp., a newly described gastrotrich from the northern Gulf of Mexico. This marine species inhabits fine sediments with little detritus.

Scanning electron micrograph by A. Todaro.

✓ Checklist of Suggested Demonstrations

THE PSEUDOCOELOMATES

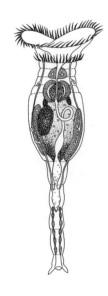

_____ 1. *Ascaris* preserved specimens.

_____ 2. *Dirofilaria* (dog heartworm) preserved specimens in heart.

_____ 3. *Ancylostoma* (hookworm) whole-mount slide.

_____ 4. *Philodina* whole-mount slide and accompanying diagram.

_____ 5. *Philodina* living specimens in culture.

_____ 6. *Lepidodermella* whole-mount slide and accompanying diagram.

_____ 7. *Lepidodermella* living specimens in culture.

_____ 8. Horsehair worm preserved specimens.

Pseudocoelomate Notes and Drawings

Pseudocoelomate Notes and Drawings

THE EUCOELOMATE WORMS

THE ANNELID WORMS

Annelids are cephalized, bilaterally symmetrical worms whose bodies possess a spacious coelom, the **eucoel**, which is fluid-filled and fully lined by mesodermal tissue. This group includes the earthworms, the leeches, and a variety of other worm-like organisms. The annelids are well equipped for aquatic life and occur in freshwater, marine, and even damp terrestrial situations. They are abundant in all of these environments but they are most diverse in marine habitats. They may be free-living or parasitic. Annelids have many adaptive features and are quite advanced in comparison with the "lower" worms previously discussed in this chapter. They are divided into serial, repetitive body segments termed **metameres**. The internal organs, suspended in eucoel, are serially repeated to correspond to the external segments. Certain groups of segments are modified and specialized to perform specific functions. These groups are called **tagmata;** they result from the evolutionary fusion of individual segments to increase functional efficiency. The annelid body architecture is similar to the tube-within-a-tube layout of the nematodes, but they possess a highly developed muscular system that allows more complex movement, both internally and externally. The gut is a straight tube supplied with its own musculature, so that it functions independently of muscular activity in the body wall. The excretory and circulatory systems are well developed, and some members of the phylum even have respiratory organs. The nervous system is complex and concentrated anteriorly into one or several **cerebral ganglia.** From these a **ventral nerve cord** arises with segmental ganglia.

The Annelida is one of just a few phyla that exhibit marine, freshwater, and truly terrestrial species. Their great diversity in form and habitat attests to the adaptive success of the phylum and the worm body plan as well. Annelids swim and crawl with equal agility, they form well-constructed tubes and burrows and many display elaborate patterns and vivid colors.

SYSTEMATIC OVERVIEW

PHYLUM ANNELIDA

The segmented worms; approximately 15,000 species.

Class Polychaeta

Polychaetes are mostly marine with a few freshwater species; typically numerous setae are found on segmentally arranged parapodia and a well-developed head (*Aphrodite, Arenicola, Chaetopterus, Nereis, Sabella*).

Class Oligochaeta

Oligochaetes are found in the terrestrial, marine and freshwater environments; no parapodia exist and they possess few setae; head reduced; monoecious, with a clitellum that secretes an egg cocoon (*Lumbricus, Tubifex*).

Class Hirudinea

Leeches are found in freshwater marine and the terrestrial environments; no parapodia or setae; anterior and posterior suckers; no internal segmentation except in the nervous system; coelom reduced to network of small spaces by musculature and connective tissue; monoecious with a clitellum (*Hirudo, Haemadipsa, Glossiphonia*).

PHYLUM SIPUNCULA

Sipunculans are marine, bottom-dwelling deposit feeders, with an extensible proboscis; 350 species (*Dendrostomium, Sipunculus, Themiste*).

PHYLUM ECHIURA

Spoonworms are marine, bottom-dwelling deposit feeders with a spatula-like proboscis; 100 species (*Bonellia, Urechis*).

PHYLUM POGONOPHORA

Beardworms are long, thin, gutless worms of deep oceans; the anterior end bears one or many long, filter-feeding tentacles; 100 species (*Riftia*).

CLASS OLIGOCHAETA: *LUMBRICUS*

Oligochaetes are known from marine and freshwater habitats, but they are most conspicuous as terrestrial organisms. Their average size is 10–30 centimeters in length but large individuals have approached 3 meters. Earthworms are abundant in alkaline soils that are rich in decaying organic matter. They burrow their way through the soil, swallowing earth and digesting the microorganisms within it. The are ecologically important in the cycling of many soil nutrients and their constant burrowing helps aerate the ground. Water content is also important to the distribution of terrestrial oligochaetes. Soils with a high water content, or soils that are temporarily water-logged, impair the respiratory capabilities of the worms. Worms must then rise to the surface or suffocate. In areas of optimal soil conditions, a given tract of land may support more earthworms, per gram, than cattle or horses! The freshwater and marine species either swim about freely, crawl over plant surfaces, or burrow into benthic muds. They are typically intertidal or shallow water organisms but a few species, such as *Tubifex*, live in the low oxygen waters of deep lakes and ponds. *Lumbricus,* the "night-crawler," is a typical terrestrial worm with over 100 individual segments.

PRESERVED SPECIMENS

External anatomy. Obtain a preserved specimen of *Lumbricus,* a common earthworm (Figure 2.10). Notice that a section of the body is smooth and enlarged. The individual segments here are indistinct. This is the **clitellum,** a conspicuous glandular region responsible for the secretion of protective egg cocoons. The anterior end is the region which has fewer segments relative to the clitellum than the opposite, or posterior, end. The anterior end houses the subterminal **mouth** but the head is reduced and lacks any appendages. The preoral segment projects over the mouth in a lip-like form. This is called the **prostomium;** it immediately precedes the segment which surrounds the mouth, the **peristomium.** Locate the **anus** at the posterior end. Run your finger in both directions along the surface of the worm. In one direction it feels smooth, in the opposite it is rough and scratchy. The rough texture is due to the presence of bristles called **setae** (or **chactae**). These provide traction in locomotion and are pointed backwards to maximize efficiency of movement. Oligochaetes possess only a few of these chitinous bristles per segment.

Note the coloration difference between the dorsal and ventral surfaces. The dorsal surface is darker than the ventral surface. This is a common pattern, termed **countershading,** among vertebrates and higher invertebrates, particularly aquatic species. By interplaying with the sunlight striking the animal's surface, the angle of reflection, and the background light, countershading is an effective method of camouflage.

The body of *Lumbricus* is covered with a **cuticle,** a thin protective layer. Peel off a small piece by gently rubbing the worm with your finger or a paper towel. Examine the cuticle under the dissecting microscope.

Composed of collagenous fibers, it strengthens and protects the body and prevents desiccation. The cuticle is secreted by the **epidermis** that underlies it. Note the tiny striations in the cuticle; these refract light in such a way that iridescence is produced when viewed at the proper angle.

Located along the surface of the body are several sets of **pores** which communicate with the reproductive and excretory systems. A pair of minute excretory pores, the **nephridiopores,** are found along the ventral surface of every segment, except the first three and the last. Reproductive openings are also located along the ventral surface. Segments 9 and 10 bear the openings to the **seminal receptacles,** an opening into the oviduct is found on segment 14, and segment 15 bears **male genital pores.**

Internal anatomy. Choosing a region near the posterior, carefully remove the cuticle, and observe the epidermis. It is only a single-cell-layer thick. Just underneath the epidermis are two muscular layers, the outer **circular muscles** and the inner **longitudinal** muscles. These layers are better seen in a prepared cross-section slide. Place the worm in a dissection pan and pin the worm, dorsal side up, through the extreme anterior and posterior ends, so that the worm is linear and fairly taut. Starting at the anterior end, lightly cut through the body wall until the clitellum is reached. Be careful not to cut too deeply, for if you do, you will cut through the intestinal tract and expose the soil within, thus fouling the dissection. Most of the important reproductive, digestive, and circulatory structures are found within the anterior most one-third of the body, so there is no need to continue the dissection past the clitellum. Gently pull back the body wall laterally and place small pins at regular intervals to hold the body wall back. As you do so, note the transverse **internal septa** that compartmentalize the coelom (Figure 2.11). To execute a good dissection you will need to carefully separate the septa from the body wall, noticing that the internal organs are well-developed and sit in a spacious coelom.

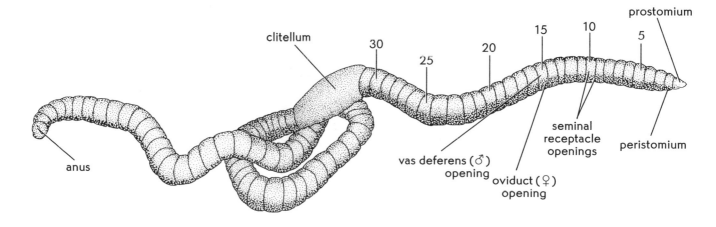

FIGURE 2.10 External anatomy of *Lumbricus,* a terrestrial oligochaete.

Digestive system. The digestive tract is a linear tube which extends from the anterior **mouth** to the posterior **anus.** It is a **complete gut** with openings at each end and specialized regions in between. The adaptive advantages over an incomplete system stem from the one-way passage of food through the gut. Without the restraints of two-way movement of material through the gut, a new meal can be engulfed before digestion of the previous meal is finished. Also, certain regions of the gut can become highly modified to perform specific functions, such as storage, grinding, and absorption. Starting with the mouth trace the digestive tract through the body and locate the structures indicated on Figure 2.11.

The mouth opens into the **buccal cavity,** but unless a meticulous dissection has been performed, you will probably see the muscular **pharynx** first. The pharynx is bound to the body wall by many fibrous **dilator muscles** which help draw food into the digestive tract. A thin-walled, tubular **esophagus** arises from the pharynx and passes to the **crop.** The esophagus is located between segments 6 to 13 but is obscured by the surrounding circulatory system. The crop (segments 14–16) is a holding structure that prepares the food for subsequent grinding and digestion. The walls of the crop are so thin that often you can see the darkness imparted by ingested soil right through the wall! It is followed by the thick-walled, muscular **gizzard** (segments 17–19) which grinds food items into small pieces to facilitate digestion and absorption in the **intestine.** The intestine continues through the remainder of the body, dumping any undigested wastes out of the body through the

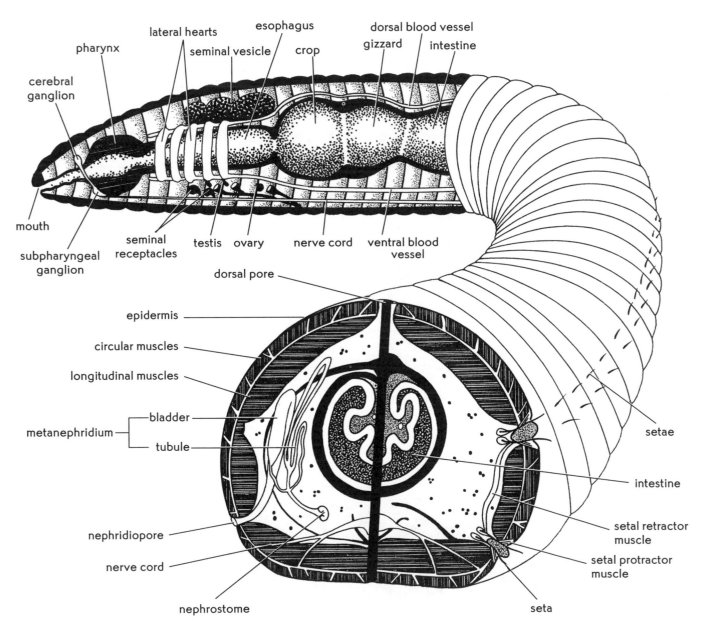

FIGURE 2.11 *Lumbricus.* Dissected specimen showing anterior internal anatomy and an offset cross-section depicting setae and a metanephridium.

anus. Notice the intestine and the dorsal blood vessel are covered with a yellowish layer of **chlorogogue cells.** Although not directly involved in digestion, this tissue stores glycogen and lipids, provides some systematic chemical defense, and generally functions as a liver analog. Lateral to the esophageal walls are several pairs of indiscrete **calciferous glands.** Designed to regulate levels of calcium and carbonate ions dissolved in the bloodstream, these excretory organs influence the pH of the blood. They develop as outpocketings of the esophageal wall but they may be difficult to see. If available, use a model earthworm to identify the esophageal glands.

Circulatory system. In keeping with the advanced features of this phylum, the earthworm has a **closed** circulatory system which contains **hemoglobin,** an iron-based pigment. The hemoglobin imparts a red color to the blood and functions as a molecular binder to transport respiratory gasses (O_2 and CO_2). The hemoglobin is dissolved within the blood fluid, termed **plasma.** The plasma is the transportation medium for the hemoglobin as well as numerous colorless **blood corpuscles.** The circulatory system consists mainly of two large blood vessels and the smaller vessels which connect them. The **dorsal blood vessel** lies on top of the digestive tract, while the **ventral blood vessel** lies just underneath it.

Smaller lateral vessels, termed **parietal vessels** communicate with the two main vessels in each body segment. However, in segments 7 to 11 the parietal vessels are greatly enlarged, muscular, and serve as pumping regions to propel the blood through the system. The 5 pairs of parietal vessels in this region are quite conspicuous and are often termed **lateral hearts,** although a more correct term would be the **aortic arches.**

Excretory system. The excretory system is tubular and organized into repeating units consistent with the segmentation of the animal. Each segment, except the first three and the last, contains a **nephridium.** Actually, each nephridium occupies two adjacent segments. The funnel-shaped opening, the **nephrostome,** extends into the previous segment and collects wastes from the coelomic fluid. The nephrostome passes posteriorly through the septal wall and connects to a coiled tubular portion which expands into the **bladder.** The bladder holds the nitrogenous waste until expulsion from the ventral **nephridiopore** is accomplished (Figure 2.12). Remember, the **calciferous glands,** discussed with digestive system, above, are excretory organs.

Nervous system. Relative to less advanced life forms, earthworms have a complex nervous system which is well equipped with processing centers ("brains") termed

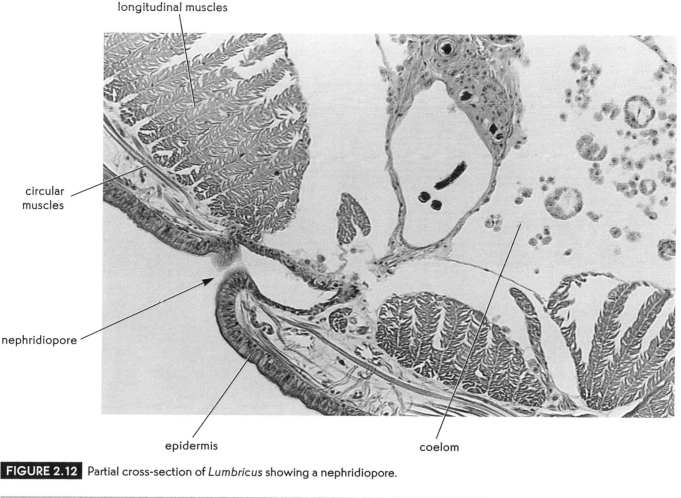

longitudinal muscles

circular muscles

nephridiopore

epidermis

coelom

FIGURE 2.12 Partial cross-section of *Lumbricus* showing a nephridiopore.

ganglia. The anterior end of the organism bears a pair of large, dorsal **cerebral ganglia,** two connecting **circumpharyngeal nerve fibers,** and a pair of ventral **subpharygeal ganglia.** All of these structures are difficult to see in a dissection, but careful preparation and a dissection microscope will reveal them. If available, a model of the dissected specimen will indicate them much more clearly. Extending from this sensory center, the **ventral nerve cord** is quite conspicuous as it runs the entire length of the body. Smaller **lateral nerves,** both **septal** and **interseptal,** ramify from the ventral nerve cord in each segment.

Reproductive system. The reproductive structures of *Lumbricus* are easily visible. Each individual worm is an **hermaphrodite,** bearing both male and female reproductive structures. Despite their **monoecious** construction, most individuals copulate to ensure cross-fertilization and genetic diversity. Typically, the male structures are larger and more conspicuous than their female counterparts. Three pairs of large, dorsal, cream-colored **seminal vesicles** store sperm until copulation. Embedded within the seminal vesicles are two pair of small **testes.** In addition, two pairs of small, bilateral, ciliated **sperm funnels** connect to two **vas deferentia** (sperm ducts) which exit the body through the male genital pore on segment 15.

The female reproductive structures are small and difficult to see. Segment 12 contains a pair of ventral ovaries which release **ova** (eggs) into the coelom. The eggs are collected by two ciliated **egg funnels,** then they travel through the **oviduct** to the **oviduct openings** on segment 14. Two small pairs of **seminal receptacles** lie ventrally in segments 9 and 10.

LIVING SPECIMENS

Locomotion. Place a live earthworm on a moist paper towel. Notice the muscular contractions (**peristalsis**) that precede forward movement. The peristaltic wave is produced by the action of the longitudinal and circular muscles upon the incompressible coelomic fluid of the hydraulic skeleton. The passage of peristaltic waves proceeds from anterior to posterior; this feature can be easily observed by making a short longitudinal slit in the dorsal body wall, and noting the opening and closing of the slit as the animal crawls. Note whether the animal moves in a particular direction and its reaction to barriers of high intensity light, ice cubes, and pools of water. Try the same experiment on a glass or very smooth surface. The animal will attempt to move but experience difficulty due to lack of traction. The earthworm has four pair of small bristle-like **seta (chaeta)** on each segment which project from the side of the body wall and impart traction during locomotion. However, this mechanism is only functional when the side of the worm is in contact with a surface (such as a burrow); when the worm is placed on a flat, smooth surface there is nothing to contact the setae and hence traction cannot be generated.

As previously mentioned, the body of the earthworm is internally partitioned into segments by septa. During locomotion, the septa act as bulkheads so the volume of fluid within each segment remains constant. When the longitudinal muscles of a segment contract, the circular muscles relax, and because the coelomic fluid is incompressible the segment becomes shorter and fatter. At the same time the setae are protruded and anchor the worm. When the circular muscles contract, the longitudinal muscles relax and the segments involved become long and thin; the setae are withdrawn, the body extends, and the worm moves forward. Because of the septa, each segment may contract or expand independently of the neighboring segments. The ventral nerve cord and the segmental ganglia control coordination between the segments.

The role of the coelomic fluid during burrowing can be demonstrated by placing the earthworm on several inches of loose, damp soil and noting the time required for burrowing. With a clean scalpel or razor blade, slice through the posterior third of the worm and separate this from the body. Note the time it takes to burrow and compare this with the trial for the intact worm.

Class Polychaeta: *Nereis, Arenicola,* and *Chaetopterus*

Polychaetes are highly segmented annelid worms which possess lateral appendages termed **parapodia.** The parapodial projections bear many chitinous **setae,** which like *Lumbricus,* are an aid in locomotion. The setae in polychaetes are often very well-developed and are much more conspicuous than those of oligochaetes. In some polychaete species the seta are modified to function as respiratory devices. Polychaetes show a number of contrasting features when compared with oligochaetes, namely, a well- developed **head,** the lack of a clitellum, the absence of permanent gonads, and the presence of a **trochophore** larvae. Containing 11,000 of the 15,000 annelid species, the polychaetes are by far the most diverse annelid group. Most species are marine, and they are quite abundant on the ocean floor and in intertidal muds and sands.

Nereis

Nereis, the clamworm, is a large marine annelid common along the North American Atlantic coast. A wandering, or **errant,** polychaete, *Nereis* burrows in sediment during the day and secretes mucus which binds the sand and mud to form loose tubes. At night, clamworms extend their body from the tube or completely leave the burrow to hunt a variety of small invertebrates. Clamworms have an extremely well-developed head with eyes, sensory tentacles, and chitinous jaws.

PRESERVED SPECIMENS

External anatomy. Examine the body of *Nereis*, the clamworm (Figure 2.13). Compare this with the structure of *Lumbricus* and note the presence of the parapodia on each segment. The parapodia are thick and fleshy and bear numerous long setae. Note the arrangement of the setae and their prominence in comparison with the oligochaete setae. The parapodia are **biramous**

(two-branched) with a dorsal **notopodium** and a ventral **neuropodium**. The anterior end of *Nereis* is occupied by a **prostomium,** a triangular tissue mass above the mouth, and the **peristomium,** the first true body segment. **Tentacles,** paired **palps,** and four **eyes** are located on the prostomium, while four pairs of tentacular **cirri** extend from the peristomium. The **pharynx** is eversible and may be distended. If so, do not confuse it with the external head structures. Two powerful **jaws** are used in prey capture, and the soft-bodied prey is held by **teeth** (denticles) in the pharynx.

Reproduction. *Nereis* is a **dioecious** organism and therefore, the gonads are only temporary structures formed during the breeding season. The **gametes,** both

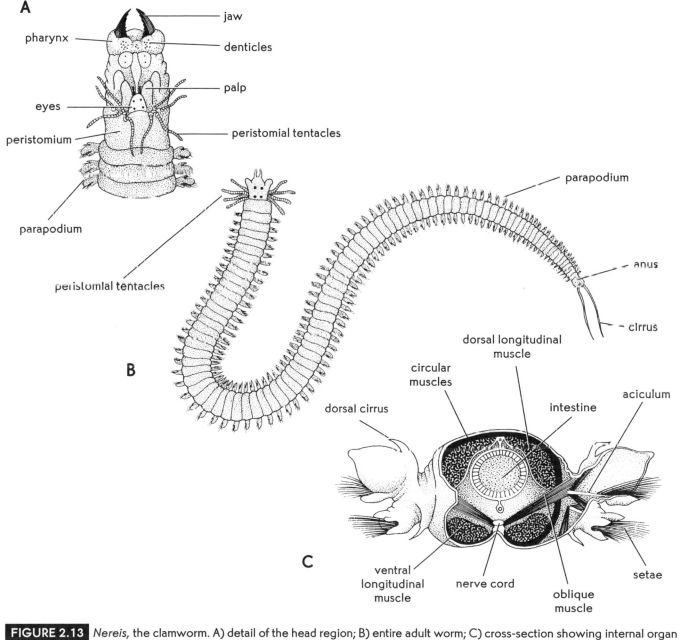

FIGURE 2.13 *Nereis,* the clamworm. A) detail of the head region; B) entire adult worm; C) cross-section showing internal organ arrangement and parapodial structure.

sperm and egg cells, are formed in the body wall and released into the coelom, and finally into the open sea. Here, external fertilization occurs and the zygotes develop into **trochophore larvae.** Depending upon the species, subsequent larval forms may develop or metamorphosis may proceed transforming the organism into an adult. This trochophore larvae is very similar to the larval forms of molluscs (clams, oysters, snails, etc.) and suggests a close evolutionary relationship between these two groups. Many polychaetes show remarkable behavioral and structural changes during the breeding season. Some species aggregate in huge masses and release their gametes in a sexual frenzy. This behavior greatly increases the random chance of reproductive success that external fertilization dictates.

 LIVING SPECIMENS

Locomotion. If available, obtain a living nereid for observation. First, place the worm in a dish of seawater. The clamworm is an accomplished swimmer and uses undulation and movement of the setae to produce its motion. Remove the clamworm from the dish and place it on a piece of wet paper toweling and observe the body undulate in serpentine fashion. Note the flexures of the body and successive positions of the parapodia. The clamworm can move about on land but is much more efficient in water.

Arenicola

Arenicola, the lug worm, is a burrowing polychaete which secretes a mucus-lined tube in the sediment (Figure 2.14). The lug worm is an ecological equivalent to *Lumbricus* as it burrows in the ocean sediments by extending its proboscis into the sand or mud and retracting it filled with sediment. The sediment contains detritus and plankton which are digested by the worm. Undigested sand and inorganic material pass out of the digestive tract. Relative to other polychaetes, *Arenicola* shows some structural diversity. It has reduced parapodia, a large bulbous proboscis, external gills, and reduced cephalization. If available, observe a preserved specimen and compare it with other polychaetes.

Chaetopterus

The tube-dwelling parchment worm, *Chaetopterus,* is perhaps the most bizarre annelid (Figure 2.14). The animal lives in a U-shaped tube, buried in the sand, with only the outermost ends of the tube projecting from the substrate. The animal has a variety of modified parapodia which create and move water currents, bearing food, into its tube. Observe a preserved *Chaetopterus,* and its tube, if available. Note the fan-like parapodia of segments 14, 15, and 16 which create the current. Segment 12 bears a pair of wing-like **notopodia,** which hold out a mucus bag spun by the worm.

The bag acts as a sieve; small particles are trapped in it and periodically the notopodia release the open end of the bag; it is then rolled up in the cupule anterior to the fans and transferred to the mouth by a ciliated middorsal groove. Observe that all the notopodia of the middle section of the worm are modified and adapted for spinning the bag, maintaining the water current, and handling the bag of particles after collection.

Other Tube-dwelling Polychaetes. Sabellid and serpulid worms are beautiful polychaetes with unique feeding adaptations. Feather-duster or fanworms (*Sabella*) live on the seafloor in a permanent tube constructed of sand grains cemented together with mucus. Modified head tentacles can be extended from the tube in a colorful, fan-like display used in filter-feeding. Serpulid worms are similar but form calcium carbonate tubes cemented to rocks, shells, or other hard surfaces.

Class Hirudinea: *Hirudo*

Leeches are common freshwater worms which show a secondary regression of typical annelid characteristics. They have reduced cephalization and segmentation, and lack setae altogether. They are flattened dorsoventrally and constricted by superficial rings called **annuli.** These annuli resemble external segments but, in reality, all leeches have 33 or 34 segments and these can only be identified by counting the internal arrangement of nerves. Leeches are commonly regarded as ectoparasitic bloodsuckers, and indeed many species utilize this mode of existence. However, many of the 1,000 species of leeches are free-living scavengers or predators, feeding on a variety of snails, insect larvae, and even other annelids. Most leeches live in the shallow margins of lakes, ponds, and slow-moving streams. There are some marine species and even a few terrestrial forms. They are relatively large annelids, reaching length of 30 cm. Most adults, however, are 2–5 cm in length. Parasitic leeches attach to the body surface of their hosts and suck blood or hemolymph through an incision in the body wall.

Leeches obtained a horrific reputation through several centuries of "bloodletting." Before the nature of blood formation and circulation was fully understood, man falsely believed that many diseases and bodily dysfunctions, such as fever, were caused by an excess of blood. The prime remedy for these disorders was to remove blood from the system to remove poisons or reduce pressure in the body, and the leech was a perfect vector for this methodology. Obviously, many of these medieval techniques were quite flawed and resulted in the elaboration of misconceptions regarding leeches. Today, "medicinal" leeches are useful in removing blood from bruised skin, and are important in starting and maintaining blood flow in surgically reattached limbs and digits.

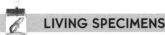

External anatomy. Examine the external anatomy of the European "medicinal" leech, *Hirudo*. The more pointed anterior end of the leech bears a small ventral **sucker**, which surrounds the **mouth**, and dorsal **eyes.** The posterior end bears a large disc-like ventral **sucker**. The suckers are utilized both in feeding and in locomotion. Note the presence of the **annuli**, and observe the pigmentation spots which serve to camouflage the organism. The **anus** is, of course, located posteriorly.

Locomotion. In leeches septa are absent, and the entire body functions as a unit. The coelom is reduced and the coelomic fluid is to a large extent replaced by tissue. The tissue is deformable and can function as a hydraulic skeleton. Leeches creep about by alternately extending and contracting, using the suckers at attachment sites. They swim actively by undulatory motions generated by rhythmic muscular contractions. Obtain a living leech, drop it into a container of freshwater and observe the swimming reaction. Place a leech on a glass plate and watch its movements; watch the position and action of the suckers.

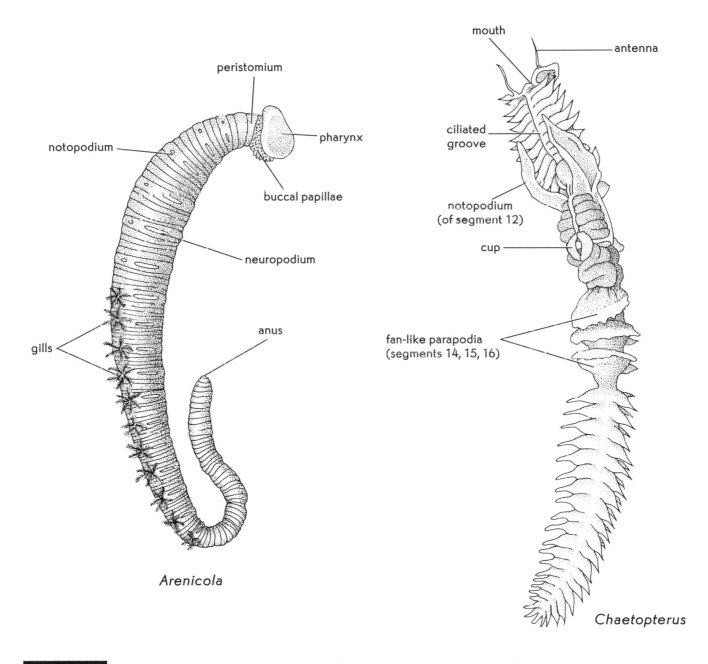

Arenicola

Chaetopterus

FIGURE 2.14 Tube-dwelling polychaetes, *Arenicola* and *Chaetopterus*, showing diversity of form.

✓ Checklist of Suggested Demonstrations

THE ANNELIDS

_____ 1. Earthworm cocoons.

_____ 2. *Lumbricus* preserved specimens.

_____ 3. *Lumbricus* cross-section slide.

_____ 4. *Lumbricus* living specimens.

_____ 5. *Nereis* preserved specimens.

_____ 6. *Nereis* cross-section slide.

_____ 7. *Chaetopterus* preserved specimen and tube.

_____ 8. *Arenicola* preserved specimen.

_____ 9. *Aphrodite* preserved specimen.

_____ 10. *Sabella* preserved specimen.

_____ 11. Polychaete tubes.

_____ 12. Preserved leeches.

_____ 13. *Hirudo* preserved specimen.

_____ 14. Living pond and medicinal leeches.

Annelid Notes and Drawings

Annelid Notes and Drawings

Annelid Notes and Drawings

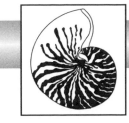

EXERCISE 3

The Molluscs

THE MOLLUSCS

SYSTEMATIC OVERVIEW

PHYLUM MOLLUSCA

A diverse phylum of soft-bodied organisms encompassing 110,000 living and extinct species, many of which possess an external calcareous shell.

Class Aplacophora

Solenogasters and caudofoveates; worm-like molluscs without a well developed mantle or shell (*Chaetoderma, Neomenia*).

Class Monoplacophora

Single-shelled marine species with several pairs of gills and nephridia; most extinct but several species live in deep ocean water (*Neopilina*).

Class Polyplacophora

Chitons; shell consists of eight dorsal plates (*Chiton, Crypto-chiton, Katharina*).

Class Bivalvia

Clams and their relatives; branchial filter-feeders which lack a radula; mantle is bilobed and produces two lateral shell valves (*Anodonta, Crassostrea, Mercenaria, Mya, Mytilus, Ostrea, Pecten, Teredo*).

Class Gastropoda

Snails and related organisms; most are asymmetrical because their body undergoes torsion during development; freshwater, marine, and terrestrial (*Aplysia, Busycon, Dendronotus, Diodora, Haliotus, Helisoma, Helix, Murex, Stramonita (Thais)*).

Class Scaphopoda

Tusk or tooth shells; possess a conical shell open at both ends; lack a distinct head (*Dentalium*).

Class Cephalopoda

Squids, octopuses, nautiluses; predaceous marine organisms with a well-developed head often possessing a sharp beak to inject poison; shell is present, reduced, or entirely lacking (*Loligo, Nautilus, Octopus, Sepia*).

The Molluscs are a group of incredibly diverse soft-bodied animals. Encompassing at least 110,000 species, the Mollusca is the second largest phylum in the animal kingdom. The group includes such commonly known animals as the clams, oysters, octopus, squid, and snails, but also includes the more obscure nautili, chitons, and tooth shells (Figure 3.1). They are bilaterally symmetrical with well-developed digestive, circulatory, excretory, and respiratory systems. A protective calcareous shell is often present, although modified or absent in some groups. The shell grows along with the soft parts of the animal, and hence does not have to be shed and regenerated during growth. Differential growth, often promoted by seasonal shifts in conditions, produces a ringed effect on the shell. These rings wear down over time and do not always correspond to yearly fluctuations but can be used to measure age to some extent. Underlying the shell is a thin, membranous layer, the **mantle**, which is reponsible for shell secretion and some respiratory, excretory, and sensory functions. In those species lacking a shell, the mantle is highly adaptable and may be thickened and muscular for protection or locomotion. The mantle completely surrounds the organism and creates a space in between the mantle wall and the viscera. This **mantle cavity** houses **gills** (or lungs in terrestrial species) which remove oxygen from the water (or air). Because air and water currents circulate through the mantle cavity, this system allows the organism to communicate chemically with the environment while still being covered by a relatively impervious, protective shell.

Molluscs are related to annelids in their development and the presence of a **trochophore larva,** but they lack segmentation. The most advanced groups are highly cephalized, but the more primitive groups are uncephalized. In modern species, the coelom is restricted to the area surrounding the heart and to spaces within the gonads and kidneys, but embryonic and fossil evidence indicates that the coelom was spacious and prominent in ancestral species. Although the success of the phylum centers around marine species, their organ systems are highly adaptable and hence terrestrial, estuarine, and freshwater species are common.

The typical molluscan body plan consists of three major regions: the anterior **head,** the ventral **foot,** and the dorsal **visceral hump** (or visceral **mass**). Each of these three regions is elaborated or suppressed in response to evolutionary pressures, hence different groups of molluscs show great diversity in these regions. An important feature of many mollusc species is the **radula,** a unique feeding apparatus which bears

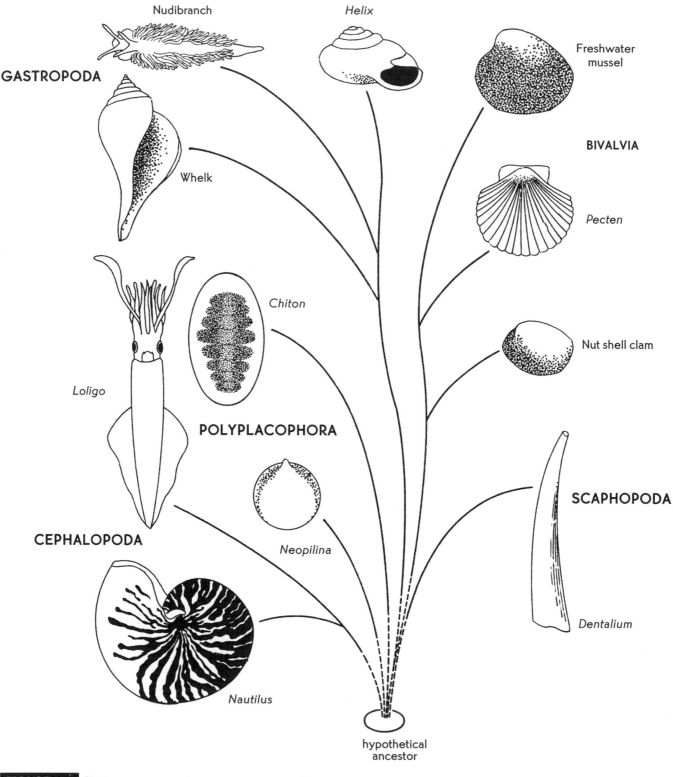

FIGURE 3.1 Phylogenetic tree of the primary classes of molluscs. Organisms pictured are representatives of each class.

Introductory Zoology Laboratory Guide

spike-like teeth and functions as a rasping organ. Often referred to as a "tongue with teeth," the radula is a tireless structure which oscillates back-and-forth over a surface, raking in organic matter, scraping up algae from rock surfaces, or, in predatory species, even grinding down other mollusc shells to expose the soft-parts.

CLASS POLYPLACOPHORA: CHITON AND CRYPTOCHITON

The polyplacophorans (amphineurans) are an ancient class of dorsoventrally-flattened, oval-shaped organisms called **chitons**. Chitons are in many ways the most primitive of molluscs, showing traces of segmentation from proto-annelid ancestors. Their fossils date back at least 500 million years, to the time when living organisms were first developing hard shells. The chiton shell consists of a unique assemblage of eight overlapping plates which are embedded in the mantle. The mantle is thick and muscular and often bears calcareous bristles or spines for protection and camouflage. The class is entirely marine, and chitons are well adapted for life along rocky tidal shores. Their low profile and muscular foot keeps them attached to the rocky substrate and prevents dislodging due to wave action. The overlapping plates and muscular mantle impart some flexibility to the animal, maximizing its ability to adhere to the uneven surfaces inherent in tidal areas. Although uncephalized, chitons have an anterior mouth which houses a scraping **radula** used to loosen algae from the rocks (see Figure 3.2, a comparison of mollusc radulae). Chitons generally average 10 cm in

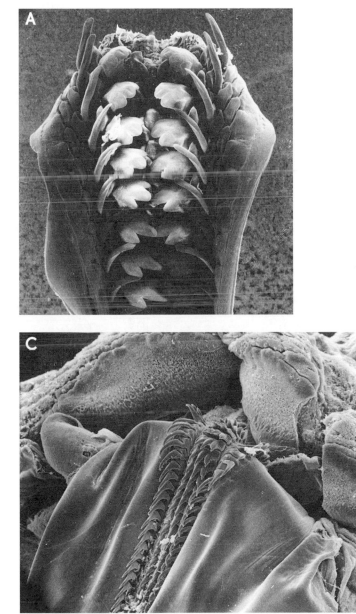

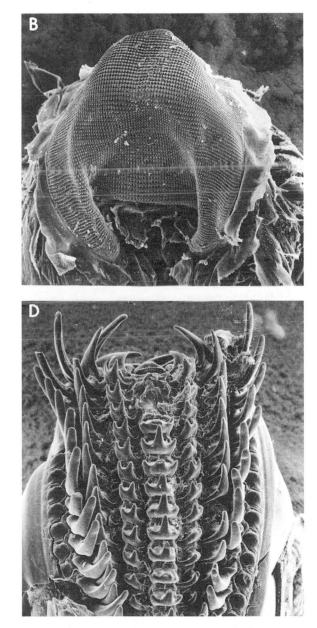

FIGURE 3.2 Various radulae of molluscs. A) *Chiton*, an herbivorous scavenger; B) *Helix*, an herbivorous snail; C) *Stramonita* (*Thais*), the oyster drill, a predaceous gastropod; D) *Loligo*, a predaceous cephalopod.

Scanning electron micrographs by R. A. Roller.

length but some species are quite large, attaining lengths of up to 30 cm.

Chiton

Obtain a specimen of either *Chiton* or *Katharina* (a similar genus) and examine the external structure (Figure 3.3). The dorsal surface is covered by the mantle; the center of the dorsum reveals portions of the eight **plates** which comprise the **shell.** Underlying the **mantle,** is the internal **visceral mass.** Turn the chiton over so that the ventral surface is visible. The muscular **foot** occupies most of the ventral surface. Between the mantle edge and the foot there is a small groove, termed the **pallial groove,** in which lie the **gills** (termed **ctenidia** in molluscs). The ctenidia are arranged in a series that runs the entire length of the body. Locate the **mouth** and **anus.**

Cryptochiton

This genus is similar to *Chiton* internally, but some external differences are obvious. *Cryptochiton* is large (20–30 cm) and appears to lack the external plates of the shell. The eight plates are present, but are completely covered by the mantle and not visible. Indeed the name of the genus means "hidden chiton," a reference to the cryptic nature of the eight plates. Examine the specimen externally and locate the ventral structures discussed for *Chiton.*

CLASS BIVALVIA: CLAMS AND MUSSELS

Bivalves, one of the largest and most ecologically significant mollusc groups, are aquatic animals housed within a hinged shell. The bivalve body is laterally compressed and surrounded by two hinged **valves**

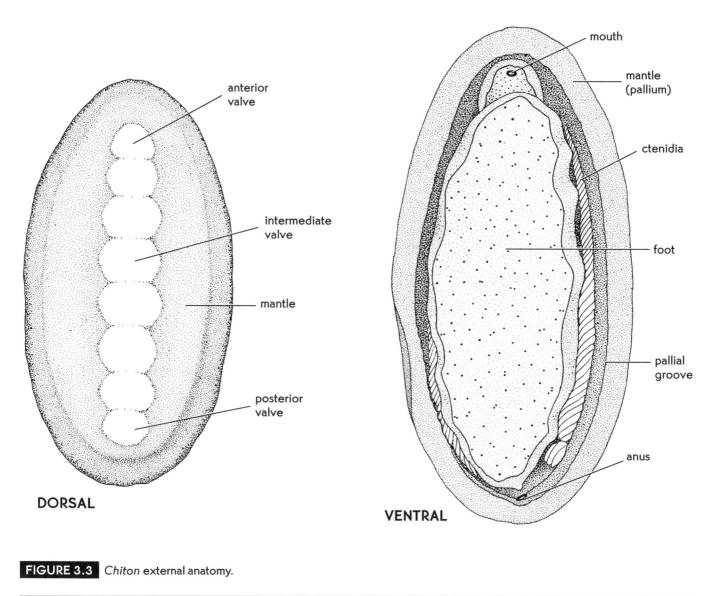

FIGURE 3.3 *Chiton* external anatomy.

which grow concentrically from a dorsal area near the hinge, the **umbo**. Bivalves are uncephalized and their sensory, circulatory, and locomotory structures are relatively primitive, yet they are quite diverse and abundant. They are filter-feeders (thus, lacking a radula) and well adapted to a burrowing existence. Most bivalves are marine organisms, a few are freshwater, and nearly all burrow into the benthic sediments by action of the muscular **foot** which can protrude from the valves and dig into the substrate. Many variations of this lifestyle exist, with some species lying completely buried within the sand or mud. These species usually have their **excurrent** and **incurrent siphons** (posterior extensions of the mantle which allow water circulation) modified to accomplish this task. Razor clams (*Ensis*) have streamlined, elongate shells and tubular siphons to facilitate deep burial. The geoduck, a freshwater mussel with fused, fleshy siphons, lives nearly 1 meter deep in mud and sand. *Tridacna*, the giant clam, sits with its hinge down and siphons exposed to sunlight, an adaptation which facilitates the photosynthetic activities of symbiotic dinoflagellates. The giant clam is famous for its weight and girth which approach 1,200 kg and 1 meter.

Some bivalve species have departed from a typical benthic existence, opting to tunnel into wood or rock. Rock-burrowers may resemble sand-burrowers, but closer inspection reveals serrations along the anterior edge of the valves. While the foot secures the animal to the rock, the serrations rub coarsely against the rock, gradually abrading it. *Teredo*, a wood-borer, possesses valves reduced to small, but sharp, cutting facets. The tiny valves open and close like jaws, forming long tunnels as the wood is continually rasped. Other bivalves have forsaken burrowing completely and attach themselves to the surface of rocks and pilings. *Mytilus* secretes fine **byssal threads** which permanently attach the animal to the substrate.

 PRESERVED SPECIMENS

External anatomy and dissection. Separating the valves of a mussel or clam can be an arduous task if not executed properly. Once the shell is removed, the dissection is fairly simple. You will need a blunt probe, a scalpel, a dissecting pan, and of course a preserved freshwater clam or mussel (any of the genera *Mercenaria*, *Anodonta*, *Mytilus*, or *Unio* will be acceptable). First, to determine the position of the internal organs, you must orient the animal properly. Hold the clam so that the **dorsal surface** (indicated by the hinge ligament and umbo) is facing you. In *Mytilus* this is the anterior portion of the animal. In *Mercenaria*, and most other genera, the umbos are more dorsally located. The **ventral surface** (where the valves open) should be pointed

away from you. Relative to the position of the umbo, one end of the clam will be longer (farther from the umbo) than the other. This "longer" end is the **posterior end,** and the "shorter" end, closer to the umbo, is the **anterior end.**

To open the valves you must slice through the two **adductor muscles** that tightly bind the valves. These muscles are quite strong and in preservative they often harden, making the opening of the valves nearly impossible unless these muscles are completely cut. The **anterior adductor** lies slightly anterio-ventral to the umbo, close to the edge of the shell. The **posterior adductor** lies near the posterior edge underneath the end of the hinge. Most commercially preserved clams are shipped with small wood blocks jammed between the valves to keep them open. Insert your blunt probe between the valves at the site of the wood block (or any spot where it will fit) and work the probe to either end of the shell. This should create enough space to insert a scalpel and carefully cut the adductor muscles. When both adductors are completely cut, the valves should open slightly. They can now be opened fairly easily by hand.

Examination of the shell reveals an irregular line on the inner surface. This marks the attachment point of the mantle. On the outside of the shell is a thin proteinaceous covering called the **periostracum**. The middle layer, which is visible in cross-section by breaking off a piece of shell, is the **prismatic** layer; it is composed of calcium carbonate ($CaCO_3$). The inner lining of the shell, the **nacreous** or "mother-of-pearl" layer, is composed of the iridescent mineral argonite. Polishing this layer produces a reflective array of colors, prompting this material to be used for modern jewelry, souvenirs, and even ancient currency. Each of these layers is secreted by the mantle. Concentric lines of growth surround the **umbo**, which is the portion of the shell that was formed first. Succeeding phases of growth and shell deposition are indicated by the concentric lines which surround the umbo.

Internal anatomy. Pulling the valves apart reveals a thin, membranous **mantle** which lines the inside of both valves and creates a spacious **mantle cavity** (Figure 3.4). The middle layer of the mantle is sensory and in some species may have eyes and/or tentacles. The innermost layer of the mantle is muscular; when the lobes of the mantle are pressed together, they form the **siphons**. Immediately visible within the mantle cavity are a pair of large ciliated **ctenidia** (gills). Each gill consists of two **lamellar folds** which are "ribbed" with invaginations to increase the functional surface area, and hence maximize gas exchange. The gills are adapted for both filter-feeding and respiration, and when water enters the mantle cavity it passes directly over the gills where food particles are trapped, gas exchange occurs, and gametes are caught during the breeding season. The ventral **foot** is a muscular extension of the **visceral mass,** a housing structure for the digestive and reproductive organs.

Bivalves are mainly ciliary-mucus feeders. The ctenidium is the primary organ of food capture in these sedentary animals and also is responsible for the selection of food items. The **labial palps** surround the mouth and are also responsible for sorting the food. Food is trapped in a mucus strand and passed into the short **esophagus** and then into the **stomach.** Projecting from the stomach is the **style sac.** In fresh specimens the style sac will contain a gelatinous rod known as the **crystalline style.** This solid rod contains enzymes capable of digesting the carbohydrates that form the majority of the bivalve diet. The enzymes are released as the style rubs against the sac wall and is eroded. Most of the digestive tract is embedded in the dorsal region of the foot and visceral mass. Slice the visceral mass lengthwise to produce a longitudinal section. The green, pulpy material surrounding the stomach is the **digestive gland,** and the yellowish material is the **gonad.** The

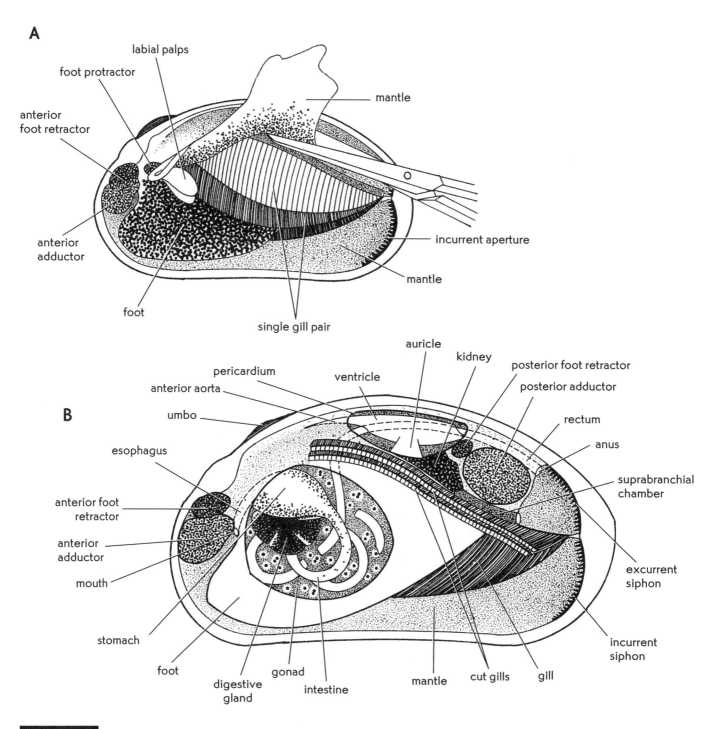

FIGURE 3.4 Clam dissection. A) removing the left mantle; B) internal anatomy.

coiled tube-like structure is the **intestine**. The intestine passes out of the visceral mass dorsally (underneath the hinge), runs through the **pericardium** and is surrounded by the tubular **heart**. The **anus** opens in the region of the dorsal, exhalent siphon. Toward the posterior end, just ventral to the heart and intestine, is a mass of brownish tissue, the **nephridia**. The nephridia is an excretory organ, similar to a kidney, which removes nitrogenous waste from the blood. In freshwater species, the nephridia is an important osmoregulatory organ, excreting large amounts of water to maintain proper water content of the body tissues.

Circulation. Bivalves have an **open circulatory system** that does not restrict the blood to closed vessels. There are some arteries and veins present to allow direct transfer of blood to and from the heart, but in the body tissues, blood passes into large cavities and sinuses. The pericardium is a thin membrane which surrounds the heart and forms the **pericardial cavity**. This cavity represents a reduced coelom and houses the thin-walled **auricles** and muscular **ventricle**. Blood travels away from the heart in two major vessels; the **anterior aorta** supplies the foot and visceral mass, and the **posterior aorta** delivers blood to the mantle and the rectum. From the viscera, blood passes to the nephridia for removal of metabolic wastes, then to the ctenidia for gas exchange, and finally back to the heart through several veins. The mantle also serves as a respiratory organ and oxygen-rich blood from the mantle returns directly to the heart.

Reproduction. Most bivalves are dioecious, although it is quite difficult to tell the sexes apart. Other species may be hermaphroditic, either protandrous or synchronous. Breeding females of some species can be easily identified by the swollen nature of the gills which function as a **marsupium**, or breeding pouch. Since most clams and mussels shed their gametes into the water, fertilization is external.

Embryonic development produces free-swimming larvae which metamorphose and settle into the substrate. Freshwater mussels depart from this method by forming **glochidia**, larval parasites which occupy the gill chambers of many species of fish (Figure 3.5). After a period of time the glochidia drop out of the gill chamber and settle into the benthos, eventually developing into mature adults.

Development. The **zygote** (fertilized egg) forms in the water column and divides into 2, 4, 8, and 16-cell embryonic stages (Figure 3.6). Cleavage in clam and other molluscan eggs is spiral and determinate, similar to cleavage in annelid eggs. The free-swimming larva is at first a **trochophore**, which later develops a velum-like fold and becomes the characteristic molluscan **veliger** larva (Figures 3.7 and 3.8). During the veliger stage in gastropods (snails) the larva undergoes **torsion**, a 180° twisting of the head and visceral mass. This is a protective larval adaptation which enables the headparts to be withdrawn into the shell behind the foot. To gain some appreciation of the developmental events and the larval forms, examine a series of prepared slides of the development showing cleavage, blastula, gastrula, trochophore, and veliger stages. Remember that the trochophore larva is also found in annelids and links these two groups phylogenetically.

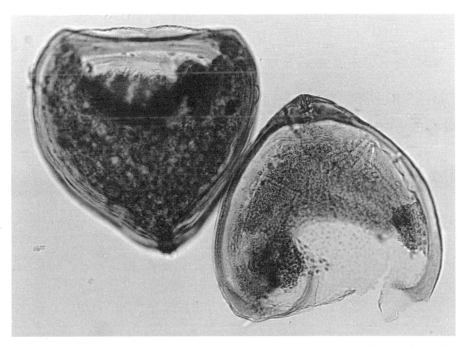

FIGURE 3.5 Glochidia, parasitic larvae of freshwater clams. These larvae occupy the gill chambers of certain species of fish before settling out into the substrate.

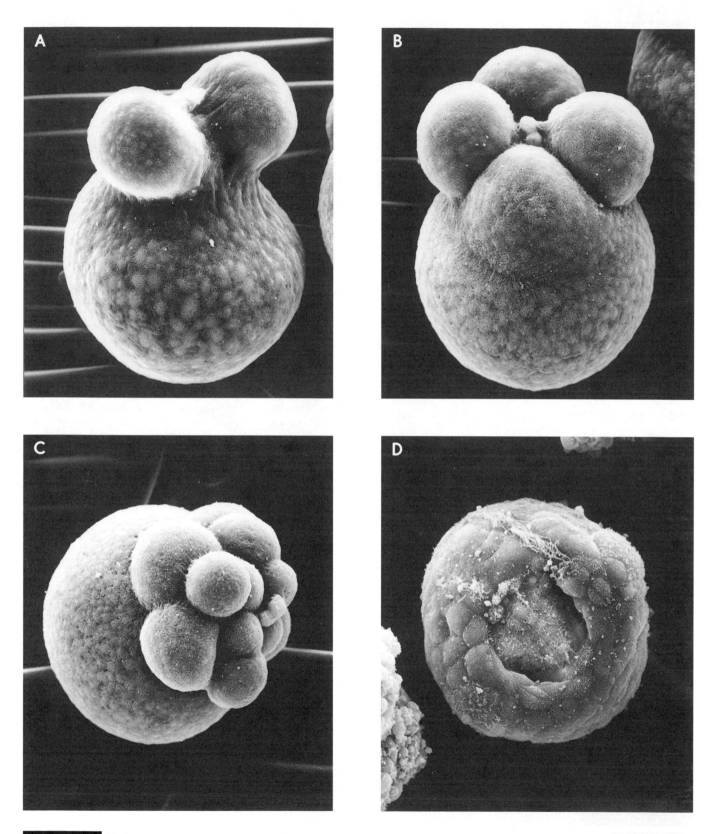

FIGURE 3.6 Molluscan developmental stages. A) two-celled stage showing first cleavage of the ovum and a centrally-located polar body; B) four-celled stage; C) sixteen-celled stage; D) gastrula stage showing the blastopore (the central depression).

Scanning electron micrographs by R. A. Roller.

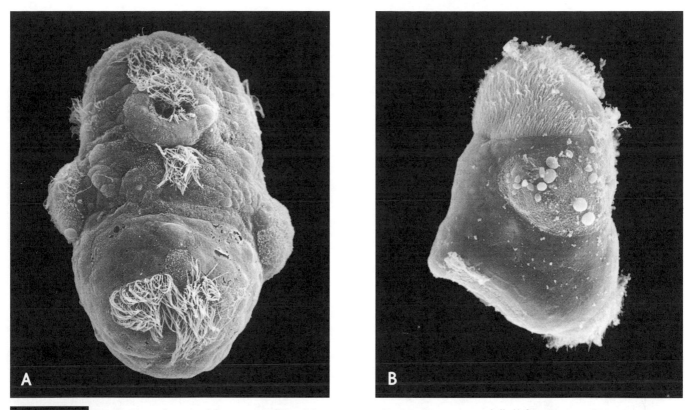

FIGURE 3.7 Trochophore larvae of *Stramonita (Thais) haemastoma*, the southern oyster drill. A) frontal view; B) lateral view.

Scanning electron micrographs by R. A. Roller.

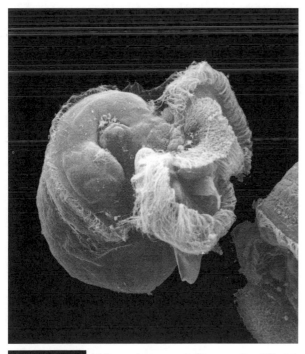

FIGURE 3.8 Veliger larvae of *Stramonita (Thais) haemastoma*.

Scanning electron micrograph by R. A. Roller.

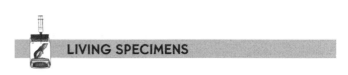

LIVING SPECIMENS

Movement. Bivalves are generally sedentary and their movements are limited. Most species employ the foot in burrowing. Place some living clams on a bed of soft sand, and watch how the foot is protruded, spread out, and anchored, and then how the rest of the body is moved by contraction of the longitudinal muscles in the foot. The scallop *Pecten* is able to swim by using its large adductor muscle to clamp shut the valves, squirting out water and propelling itself away. Being an active swimmer, *Pecten* also possesses light-sensitive eyes and sensory tentacles which protrude from the valves and are used in perception and navigation.

More typically, the adductor muscle functions in closure of the shell for protection. The adductor shortens rapidly and can maintain its tonus over long periods of time. Force a blunt probe between the valves of a live clam or mussel, and using leverage, try to open the valves. Note the strength of the adductor muscles. The interplay between the protective valves and the adductor muscles provides a great defense for bivalves. Only a few determined animals can open a closed bivalve shell, although many mammals are smart enough to break them against a rock. Starfish prey on bivalves by exerting continuous pressure on the adductors until the muscle fatigues, then digesting the accessible soft parts when the valves open.

Feeding. Observe the feeding style and respiratory mechanisms of living freshwater clams in glass aquaria. Note the presence and location of the siphons. Water currents entering and exiting the body can be easily observed by placing some carmine powder in the water near the organism. Now look for a string of mucus extending from the siphons. Glands in the gill tissue secrete mucus to trap food particles and some species produce a mucus string which is visible externally.

CLASS SCAPHOPODA: *DENTALIUM*

Scaphopods are burrowing marine molluscs found in sandy and muddy sea bottoms. The 350 species in this class are collectively called **tusk** or **tooth shells** because they possess a long, white, tubular shell which resembles an elephant's tusk. The shell is conical with an opening at each end. The larger opening is found at the anterior end which bears the **muscular foot, sensory tentacles,** and the **mouth.** Scaphopods burrow headfirst into the sand where they capture benthic microorganisms with their tentacles. Food items trapped by the tentacles pass over a toothed radula on their way to the gut. The posterior end bears a smaller opening which is usually exposed in the water column. Water, for respiration, enters and exits from this smaller opening as the foot is protracted and retracted from the opposite opening. Scaphopods lack gills but the mantle is extremely vascular and gas exchange occurs across the surface of the mantle. *Dentalium* is a typical scaphopod genus with a large tubular shell (up to 25 cm). In times past, some Native American coastal populations have used the distinctive scaphopod shell for currency.

CLASS GASTROPODA: SLUGS AND SNAILS

The gastropods are the largest and most diverse class of molluscs and the only class which has terrestrial representatives. There are approximately 40,000 living species and at least 25,000 fossil species. The fossil record is extensive for gastropods, extending back to the origin of shelled animals. This is due partly to their propensity for fossilization, because most gastropods have a well-developed shell and live in habitats that commonly produce fossils. The gastropod body plan is significantly more advanced than that of the bivalve. Usually a single shell, with a single **aperture**, surrounds and protects the organism. A number of gastropods (sea hares, slugs, etc.) have secondarily lost the shell and have no protective covering. The shell, when present, is usually **spiraled** around a central axis (the **columella**). The spiraling can develop in different ways producing several different styles (Figure 3.9). Most species have asymmetrical body parts which form when twisting, or **torsion**, occurs during larval development. Torsion, as mentioned earlier, causes the visceral mass to rotate 180° above the foot and the head. Torsion provides some adaptive advantages, especially for the larvae, but delivers some disadvantages for the adults. Without torsion, only the head faces forward. With torsion, the gut and viscera are rotated so that the mantle cavity, anus, excretory pores, and gills lie over the head. The gills are easily ventilated by undisturbed water drawn in from the front of the animal, but the placement of the anus above the mouth and near the gills creates sanitary and fouling problems.

Gastropods are classified into three large subgroups based upon the position and design of the respiratory organs. **Prosobranchs** are marine species which possess gills in an anterior mantle cavity. Of course, there are a few lunged terrestrial and freshwater exceptions. Prosobranchs (abalones and limpets) can be herbivorous, feeding on algae or carnivorous predators (cowries and whelks). The foot of many prosobranchs bears a protective calcified thickening, the **operculum**, which seals off the aperture when the foot is retracted. The **opisthobranchs** are probably detorted relatives of the prosobranchs. The mantle, shell, and gills are reduced or absent, and gas exchange occurs across the

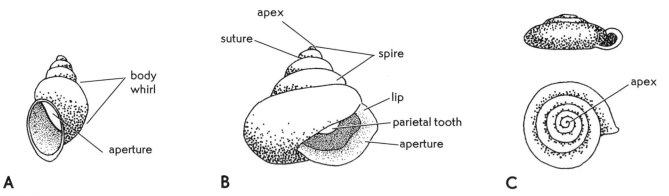

FIGURE 3.9 Structure of a gastropod shell. A) *Physa*, a sinistral (left-handed) freshwater snail with a lymnaeiform shell (height exceeds width); B) *Mesodon*, a dextral (right-handed) snail with heliciform shell (width exceeds height); C) Helicodiscus, a terrestrial snail with planospiral (flattened) shell.

body surface. Common examples are the sea slugs (**nudibranchs**), sea butterflies, and sea hares. The **pulmonates** are the most derived (advanced) gastropods. This group has left the aquatic habitats of their ancestors and invaded terrestrial environments. The mantle and the mantle cavity have been modified into an internal air-breathing lung which facilitates life out of the water. Although common on land, some pulmonates have returned to freshwater and marine environments. Terrestrial slugs are pulmonates which have either completely lost the shell or possess a small internal remnant of the shell. This may be an adaptation to reduce calcium usage in an environment deficient in calcium.

 ## SHELLS AND PRESERVED SPECIMENS

External anatomy. The **head** bears sensory **tentacles** and light-sensitive **eyes** (Figure 3.10). The **visceral hump**, which is coiled and hidden in the shell, overlies on the ventral muscular **foot**. The **operculum** is present in prosobranchs but absent in pulmonates and opisthobranchs. Obtain a preserved snail (many species are appropriate, and availability may vary with geographic location) and observe the basic external features mentioned above. Secure the specimen in your hand and, using vice grips or a hammer, crush and remove the shell from the animal. The **columellar muscle** attaches the animal to the shell and should be visible. The **mantle** surrounds the visceral mass, and anteriorly, the mantle is thick and forms a **collar** which extends around the body. The space enclosed by the mantle is the **mantle cavity.** Turn the animal to the ventral side and locate the position of the **mouth.** The **anus** and

pneumostome, if present, are located just below the edge of the shell, above the foot. The **genital opening** is located along the upper region of the foot toward the anterior.

Internal anatomy. First, dissect the head region carefully, probing the mouth for the **radula.** The radula will vary greatly depending upon the species with which you are working. Some predatory species possess a drill-like structure used to create holes in other mollusc shells. Other species may have a more generalized radula for scraping up algae and plant matter. Open the mantle cavity by making a median incision that runs dorsal and posterior toward the visceral mass. Fold back the mantle to expose the internal organs of the mantle cavity. Find the attached **ctenidium,** and anterior to it the **osphradium,** a chemoreceptor. Along the upper flap of the mantle are the **hypobranchial gland** and the **anus.** Many gastropod species are hermaphroditic. If your specimen is an hermaphrodite or a female, look for the pallial oviduct and the developing embryos. Place some embryos on a microscope slide and view under low power. You should see a variety of stages present. If your specimen is a male, you should be able to see the enlarged right tentacle which is used as a copulatory organ.

Shell diversity. Observe the great diversity of gastropod shells on demonstration. Note the unusual forms and adaptations of each. Many gastropods have modified the basic architecture of the shell for camouflage (*Murex,* the rock shells); others have reduced the spiral nature of the shell (the abalones, *Haliotus;* the slipper shells, *Crepidula;* and the keyhole limpet, *Megathura*); and still others have completely lost the shell (nudibranch "sea slugs"; *Aplysia,* the "sea hare"). Note that

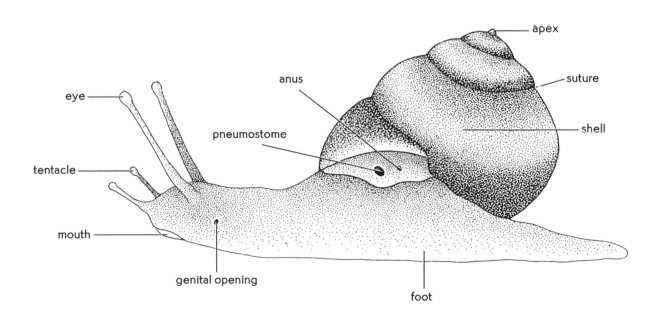

FIGURE 3.10 External anatomy of a typical gastropod snail.

some gastropod shells appear old and weathered, while others look polished. These "polished" shells are covered by the mantle in life and not exposed to the elements. When the animal dies, the shell washes up on shore and appears shiny and new.

 LIVING SPECIMENS

Movement. Slugs and many snails are active herbivores, searching their environments for suitable material to ingest. Observe a living specimen of *Viviparus* or *Campeloma* as it browses on algae-covered glass. Chitons and gastropods move by peristaltic action of the foot, somewhat similar to the crawling motion of flatworms. The waves can be seen by allowing a living animal to attach to the side of an aquarium and viewing the animal from the behind the glass. Note how the expanded part of the foot remains in contact with the glass by means of mucus secreted by the pedal glands, and how the contracted region advances forward.

Locomotor activity in gastropods is not always restricted to the sole of the foot. The role of the **columellar muscle** in retracting the animal into its shell is one example; this muscle serves other functions as well, and is particularly important in the righting reaction. Place a living snail specimen on its back and note the righting response. By contraction of the circular portions of the columellar muscle, the sole is pushed outward, and when it attaches to the substrate, the sharp contraction of the longitudinal muscles turns the animal upright.

Behavior. The sensory organs of gastropods are fairly well-developed in comparison with other molluscs. They do not reach the complexity shown by the cephalopods but far outdo the chitons and bivalves. **Chemoreception, photoreception,** and **geotaxis** are some fundamental sensory functions of gastropods which influence their behavior. Torsion results in bringing the chemoreceptor of the mantle cavity, the **osphradium,** to a forward position. This supplements the other sensory organs associated with the head, such as eyes and tentacles. Examine specimens of *Stramonita* (*Thais*) and identify the osphradium, eyes, and tentacles. Chemoreception is easily demonstrated by offering some oyster juice to live *Stramonita* (*Thais*); note their accelerated crawling behavior.

Photoreception is an important adaptive ability of many gastropod species. *Physa, Campeloma, Viviparus,* and *Stramonita* (*Thais*), among others, use light as a cue, directing themselves with reference to the light beam as if it were a compass. The eyes are normally borne on stalks, and contain a spherical lens. A double corneal layer covers the lens. Only light falling on the eye between 35 and 130 degrees of the body reaches the retina, so that orientation angles outside these limits cannot be used. You can determine whether the animals orient toward light or dark by placing some specimens in a container that is immersed half in darkness and half in light. Allow the animals to crawl in diffuse subdued light, and then brightly illuminate them with a horizontally directed beam. Geotaxis (orientation in response to gravitational forces) is controlled by **statocysts** which are usually located in the foot. Allow a snail to crawl along on a glass plate and note the change in its direction of movement as one edge of the plate is lifted, for example, the edge in front of the snail.

CLASS CEPHALOPODA: *NAUTILUS* AND *LOLIGO*

From a structural and behavioral standpoint, the cephalopods have reached the pinnacle of molluscan evolution. Their sensory structures are extremely well designed and include an eye which rivals the vertebrate eye in complexity. Their behavior can be very complicated, including integumentary color changes and the expulsion of "ink" to confuse predators. Their circulatory system is **closed,** an advancement over the open, largely unvesseled systems of other molluscs. They are a strictly marine group that includes the octopods, squid, cuttlefish, and the nautili. Cephalopods are active, highly motile swimmers that generally move head-backwards by means of water-propulsion. Water is pumped into and out of the mantle cavity by the contraction of strong muscles, forcing the water to be expelled through a tube-like **siphon.** Squid are among the fastest invertebrates, swimming at speeds of 43 km/hr. The giant squid, *Architeuthis,* is the world's largest invertebrate reaching a length of 18–20 meters. *Architeuthis* is occasionally preyed upon by sperm whales and their epic battles are sometimes evidenced by sucker scars, left by the squid tentacles, on the whales. Some of these scars are so large that the squid that delivered them are calculated to be up to 25–30 meters long; however, no specimens quite this large have ever been captured.

Cephalopods are carnivorous animals, efficiently pursuing fishes and crustaceans. Prey is captured by agile **tentacles,** which bear **suckers,** and are modifications of the foot. The number of tentacles differs within the Cephalopoda, ranging from the decapods (squid and cuttlefish, with 2 tentacles and 8 arms) to the octopods (which lack the tentacles and have 8 arms), to finally the nautili which may bear up to 90 arms. Indeed, the cephalopod foot is by far the most highly adaptive feature of all mollusc locomotory devices. As the name of the group implies, the foot is modified into a head-like analogue which bears **arms, eyes,** nerve **ganglia,** a piercing chitinous **beak,** and toxic **poison glands** which help immobilize the prey. The shell is usually reduced (an internal proteinaceous rod in squid) or absent (octopods). Many ancient cephalopods, such as the **ammonites,** were common marine animals with a thick outer shell. Today, only the *Nautilus* retains an external shell (Figure 3.11).

 SHELLS AND PRESERVED SPECIMENS

Nautilus

A primitive cephalopod representative, *Nautilus* is a remnant of a once successful group of shelled marine cephalopods that were common during the dinosaur age. Today, the nautili are restricted to the tropical Pacific, where they spend most of their time on the ocean floor searching for prey. In comparison to other cephalopods, nautili are slow swimmers, a trade-off for a thick, protective, somewhat cumbersome shell. The shell of *Nautilus* is **multichambered** and the organism occupies only the outermost chamber. As the *Nautilus* grows it secretes a new shell chamber and seals off the old chamber, leaving only a small opening in the center of the partition. A long tube of tissue, the **siphuncle,**

passes through this opening and those of the previous chambers, providing chemical and metabolic communication with the inner chambers. The siphuncle absorbs fluid from the empty chambers which subsequently fills with gas, a by-product of respiration. The animal regulates the amount of gasses in these internal chambers to increase or decrease buoyancy, and hence regulate its placement in the water column. Nautili are not found below depths of 600 meters because the water pressure at that depth is so great (over 40 times that of atmospheric pressure) that it superceeds their ability to equalize internal pressures and their shells collapse. Examine the thick calcified shell of *Nautilus* and note the distinctive color pattern. If available, observe a sectioned shell and look for the inner chambers, which are larger near the aperture. Observe the passageway in the chamber partitions that house the siphuncle.

Loligo

Squid are extremely common marine animals, often swimming near the surface in large schools. They are found in both the Pacific and Atlantic coastal waters and some species inhabit the ocean depths. Squid are

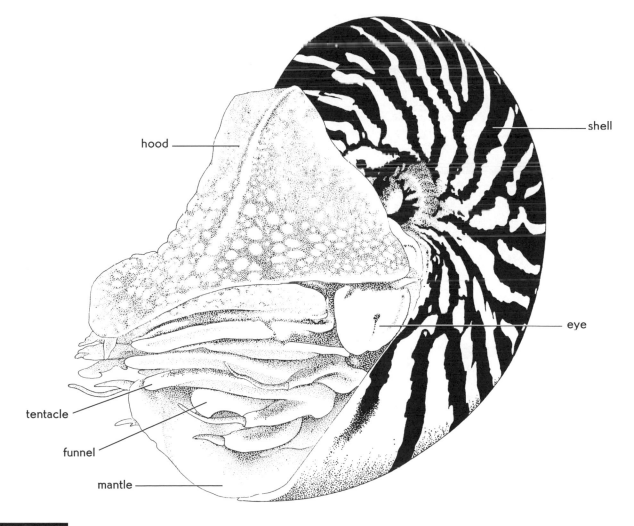

FIGURE 3.11 External anatomy of *Nautilus,* a shelled cephalopod.

favorite food items of many fish and sharks, and are an important food source in many Asian cultures. To facilitate swimming, squid are streamlined with a tapered leading end, the **apex**. The orientation of the squid is quite unusual because the animal is extended dorsoventrally. Despite resembling an anterior end, the head and tentacles represent the **ventral surface**. The leading **dorsal surface** is pointed to reduce water resistance when swimming and opposes the head and tentacles. The **funnel** or **siphon** is found along the **posterior**, while the opposite side represents the **anterior.**

External anatomy and dissection. Examine a preserved specimen of the squid, *Loligo* (Figure 3.12). Note that there is no external shell and that the major part of the body is enclosed by the muscular, fleshy **mantle.** Note the pattern of **chromatophores**, or pigment spots, on the surface of the mantle. Squid have the ability to camouflage the mantle by manipulating the surface pigmentation. Observe the **head-foot** region, and note the eight **arms** are completely lined with suckers. Two of the appendages are longer than the arms. These **tentacles** only bear suckers at the expanded distal tips and can be extended and retracted by internal muscle control. Place the animal so that the apex is farthest from you and the arms closest. Turn the animal so that the **siphon** (the tubular projection between the opening of the mantle and the head-foot region) is facing you. The **eyes** are on the left and right sides of the body, and the head has moved to a position that is dorsal to the foot.

Squid are active predators, and the walls of the mantle cavity play an important locomotory function.

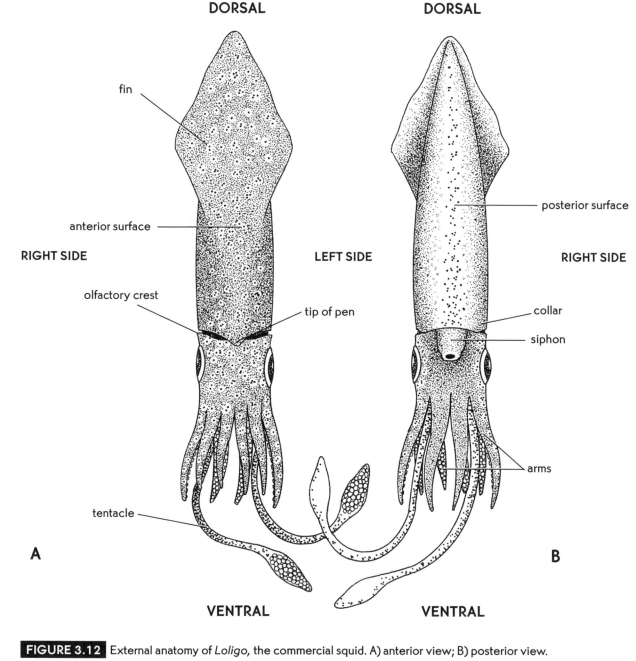

FIGURE 3.12 External anatomy of *Loligo*, the commercial squid. A) anterior view; B) posterior view.

Highly muscularized, the mantle controls movement of water within the cavity. With the ventral side of the squid facing you, observe the construction of the mantle collar and its relationship to the siphon. In life, the mantle cavity expands by muscular action, allowing water to enter. The collar then locks tightly against the head, utilizing a system of ridges and grooves in the mantle. Once the mantle collar is locked in place, the siphon is the only exit pathway from the mantle cavity. The siphon is well equipped with muscles, and can be pointed for making directed jet-propulsive movements.

Internal anatomy. Open the mantle cavity by making an incision which runs the entire length of the posterior surface from siphon to apex. Keep your scalpel close to the body wall so as not to destroy any internal organs. Turn the mantle edges laterally and insert some heavy dissection pins at frequent intervals. After preservation, the mantle may be quite rubbery and resistant to pinning, be patient! Note that the tip of the siphon has a **valve** which regulates the outflow of water from the mantle cavity (Figures 3.13 and 3.14). The inner side of the mantle has cartilaginous ridges which keep the inhalant currents separate from the exhalent. There are two **ctenidia (gills)** which are placed so that the

inhalant streams immediately pass over each, then converge and exit as a single exhalant stream. If the circulatory system is injected with latex (red and/or blue) study the relationship of the gills to the blood vessels and the hearts. Squid have three hearts, two round **branchial hearts** just dorsal to each gill and a triangular **systemic heart** located between the branchial hearts. A system of vessels and sinuses include the conspicuous anterior and posterior **venae cavae** and the **lateral mantle arteries** and **veins**. The blood flows from the systemic heart to the body via the **aorta**. Blood from the body pools in the venae cavae, passes to the ctenidia via the branchial hearts, and then returns to the systemic heart.

Reproductive system. Squid are **sexually dimorphic**, with individual males and females (Figure 3.13). Females have one or two short **oviducts** lying next to the rectum. Two large, conspicuous **nidamental glands** secrete albumin, yolk, and protective membranes around the eggs. The **ovary** lies anterior to the cecum in the apex. Males possess a well-developed **testis** which lies posterior to the cecum and produces sperm that passes through a **seminal vesicle** to a coiled **spermatophoric gland**. Here the sperm are enclosed in a **spermatophore**

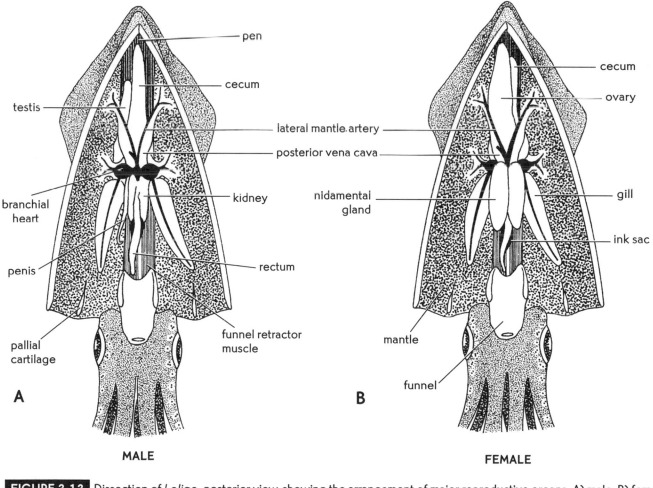

FIGURE 3.13 Dissection of *Loligo*, posterior view, showing the arrangement of major reproductive organs. A) male; B) female.

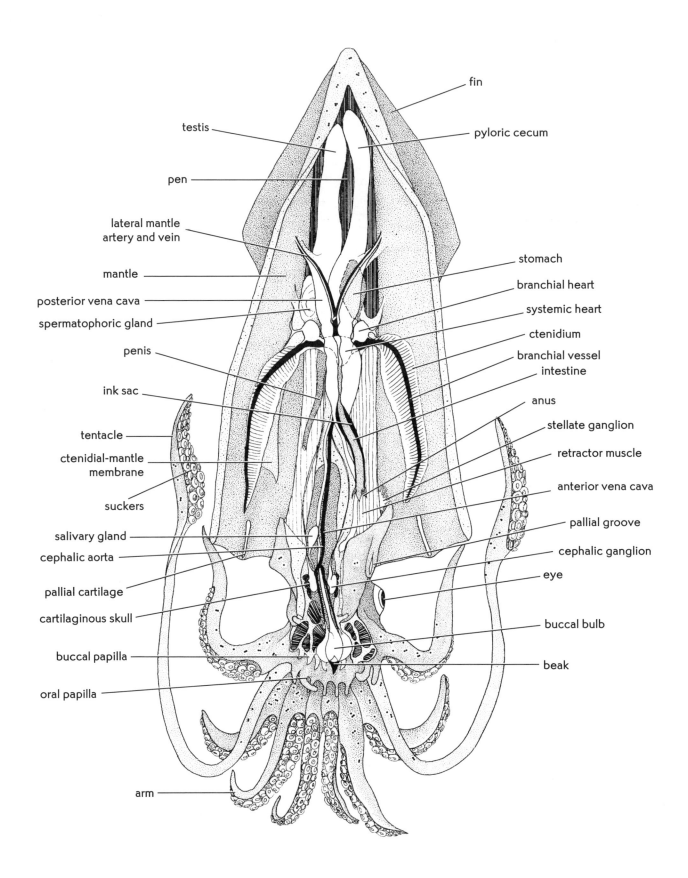

testis

pen

lateral mantle
artery and vein

mantle

posterior vena cava

spermatophoric gland

penis

ink sac

tentacle

ctenidial-mantle
membrane

suckers

salivary gland

cephalic aorta

pallial cartilage

cartilaginous skull

buccal papilla

oral papilla

arm

fin

pyloric cecum

stomach

branchial heart

systemic heart

ctenidium

branchial vessel
intestine

anus

stellate ganglion

retractor muscle

anterior vena cava

pallial groove

cephalic ganglion

eye

buccal bulb

beak

FIGURE 3.14 Dissection of *Loligo,* male specimen in posterior view, showing major internal features.

where they wait until copulation. An elongate **penis,** lateral to the intestine, carries the spermatophores to the anterior region near the funnel. In squid, the penis is not a copulatory organ.

Squid show a complex mating ritual where they encircle one another while waving their arms and tentacles and displaying color changes. During copulation, they line up head-to-head, intertwine their arms and one of the male's arms, the **hectocotylus,** reaches through his funnel, picks up spermatophores and inserts them into the female's mantle cavity, where fertilization occurs.

Digestive system. The **mouth** is ventral and surrounded by the tentacles and arms. The prey is held firmly in place by these prehensile arms and suckers, and food is brought toward the mouth. Remove the siphon, and by a superficial median incision, cut into the head, separating the eyes, and exposing the round **buccal mass** (Figure 3.14). This is a muscular organ which possesses two horny, raptor-like **beaks,** used for ripping and shredding the prey. Pry open the beak and observe the internal radula and the odontophore. Run your finger along the radula and feel the serrations. Place the radula under a dissecting microscope and compare with the scanning electron micrograph in Figure 3.2.

Posterior to the buccal mass are a pair of **salivary glands** which pour their toxic secretions into the buccal cavity as the prey passes into the mouth. These may be indistinct and hard to identify. Pull out the buccal mass and identify the thin-walled **esophagus.** The esophagus is surrounded by the liver and runs to the muscular **stomach.** The stomach emerges to form the gelatinous **cecum,** which is harbored in the apex. The **intestine** runs toward the head from the stomach and terminates in the **rectum.** The anus opens near the internal opening of the siphon, so that waste products are immediately expelled. A diverticulum extends from the intestine; this is the infamous **ink sac** that produces a murky secretion, clouding the water when expelled in an effort to deter predation. The ink is a conglomerate of pigment, which not only visually obscures the view of a predator, but may desensitize a predator's chemoreceptors.

✓ Checklist of Suggested Demonstrations

THE MOLLUSCS

_____ **1.** Trochophore larvae whole-mount slide.

_____ **2.** Veliger larvae whole-mount slide.

_____ **3.** Preserved chiton specimens and isolated valves.

_____ **4.** *Chiton* preserved specimen.

_____ **5.** *Cryptochiton* preserved specimen.

_____ **6.** *Chiton* radula whole-mount slide.

_____ **7.** Assorted bivalve shells and preserved specimens.

_____ **8.** Glochidium larvae whole-mount slide.

_____ **9.** Clam gill cross-section slide.

_____ **10.** *Pecten* (scallop or Lion's paw) shell.

_____ **11.** Angel's wing shell.

_____ **12.** *Teredo* preserved specimen and drilled wood block.

_____ **13.** Various cockle shells.

_____ **14.** Assorted gastropod shells.

_____ **15.** Snail anatomy diagram.

_____ **16.** Snail radula whole-mount slide.

_____ **17.** Conch shell cut in longitudinal section.

_____ **18.** *Cypraea* (tiger cowry) shell.

_____ **19.** *Polineces* (moon snail or shark's eye) shell.

_____ **20.** *Stramonita* (*Thais*; oyster drill) shell.

_____ **21.** Egg cases of *Stramonita* and *Busycon*.

_____ **22.** *Crepidula* (slipper shell) shell.

_____ **23.** *Haliotus* (abalone) shell.

_____ **24.** *Conus* (cone shell) shell.

_____ **25.** *Murex* (rock shell) shell.

_____ **26.** *Megathura* (giant keyhole limpet) preserved specimen.

_____ 27. *Aplysia* (sea hare) preserved specimen.

_____ 28. Nudibranch (sea slug) preserved specimen.

_____ 29. *Dentalium* (tusk or tooth shell) shell.

_____ 30. Fossil cephalopod shells.

_____ 31. *Nautilus* shells (whole and sectioned).

_____ 32. *Octopus* preserved specimen.

_____ 33. *Loligo* preserved specimen.

Mollusc Notes and Drawings

Mollusc Notes and Drawings

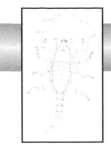

EXERCISE 4

The Arthropods

SYSTEMATIC OVERVIEW

PHYLUM ARTHROPODA

Metameric protostomes with jointed appendages, a chitinous exoskeleton, and superficial cleavage; the largest and most diverse phylum in the animal kingdom, over 1 million known species.

Subphylum Trilobita

Extinct marine arthropods; body was divided into three lobes and bore many uniform, biramous appendages.

Subphylum Chelicerata

Body subdivided into an anterior prosoma and a posterior opisthosoma; usually four pairs of walking legs; anterior appendages include chelicerae and pedipalps; no mandibles or antennae.

Class Merostomata

Marine chelicerates with abdominal gills; includes giant water scorpions (extinct) and horseshoe crabs (*Eurypterus*, *Limulus*).

Class Arachnida

Scorpions, pseudoscorpions, spiders, ticks, mites; prosoma bears four pairs of walking legs; most are terrestrial.

Order Araneae — Spiders; unsegmented abdomen is attached to the cephalothorax by a narrow waist (*Argiope*, *Latrodectus*).

Order Acarina — Ticks and mites; small, with unsegmented abdomen broadly fused with a cephalothorax (*Dermacentor*, *Sarcoptes*, *Trombicula*).

Order Opiliones — Harvestmen ("daddy longlegs"); small segmented abdomen with long legs (*Habrobunus*).

Order Scorpiones — Scorpions; possess large, pincerlike pedipalps; abdomen segmented and elongate, with a terminal stinger (*Centruroides*, *Hadrurus*).

Order Pseudoscorpiones — Pseudoscorpions; small, scorpion-like, with large raptorial pedipalps, but no stinger or whip on posterior end of abdomen (*Chelifer*, *Vachonium*).

Order Uropygi — Whip scorpions or vinegaroons; large raptorial pedipalps, with elongate whip-like tail on posterior end of abdomen (*Mastigoproctus*).

Class Pycnogonida

Sea spiders; prosoma is large compared to the minute opisthosoma; four to six pairs of extended walking legs (*Nymphon*, *Pycnogonum*).

Subphylum Crustacea

Possess biramous appendages, two pairs of antennae, and feeding appendages (a pair of mandibles, two pairs of maxillae); most are aquatic.

Class Branchiopoda

Water fleas, fairy shrimp; flattened, seta-covered appendages; used in filter-feeding; usually freshwater (*Artemia*, *Eubranchipus*, *Daphnia*).

Class Copepoda

Most have a cylindrical body; usually long antennae, and a single median eye; marine and freshwater (*Cyclops*).

Class Ostracoda

Seed shrimps; bivalve carapace encloses the body; marine and freshwater (*Cypris*).

Class Cirripedia

Barnacles; body typically covered with calcareous plates; adults are sessile; entirely marine (*Balanus*, *Lepas*).

Class Malacostraca

Shrimps, crabs, lobsters, crayfish, pill bugs, beach fleas; body divided into a head, thorax (bearing walking legs), and an abdomen; largest group of crustaceans, includes 10,000 species (*Homarus*, *Cambarus*, *Penaeus*, *Callinectes*, *Bathynomus*).

Subphylum Uniramia

Unbranched (uniramous) appendages; one pair of antennae, and feeding appendages (one pair of mandibles, two pair of maxillae); most are terrestrial.

Class Diplopoda

Millipedes; elongate, segmented, uniramians; each segment bears two pair of legs (*Julus*, *Narceus*).

Class Chilopoda

Centipedes; elongate, segmented, uniramians; each segment bears one pair of legs (*Scolopendra*, *Scutigera*).

Class Insecta

Body divided into a head, thorax, and abdomen; thorax bears three pair of legs, and in the adult two pair of wings (usually).

Subclass Apterygota — Primitive wingless insects; no metamorphosis.

  Order Thysanura — Silverfish.

Subclass Pterygota — Winged insects; metamorphosis is either simple or complete.

Division Exopterygota — Larvae with external wing pads, gradual development, no pupal stage.

  Order Odonata — Dragonflies & damselflies, aquatic larvae.

  Order Orthoptera — Crickets, grasshoppers, mantids, roaches.

  Order Hemiptera — True bugs, piercing-sucking mouthparts.

  Order Homoptera — Scale insects, cicadas, piercing-sucking mouthparts.

Division Endopterygota — Larvae with internal wing buds, metamorphosis, larvae and adult unalike, pupal stage.

Order Coleoptera — Larvae grubs, adults beetles.

Order Lepidoptera — Larvae caterpillars; adults moths and butterflies.

Order Diptera — Larvae maggots, wigglers; adults flies and mosquitoes.

Order Hymenoptera — Larvae grubs; adults bees and wasps.

RELATED PHYLA

PHYLUM TARDIGRADA

Water bears; aquatic herbivores showing evidence of metamerism and having affinities with pseudocoelomates, deuterostomes, and arthropods; nonchitinous exoskeleton molted periodically; cryptobiotic; 500 species (*Macrobiotus*, *Echiniscus*).

PHYLUM PENTASTOMIDA

Tongue worms; unsegmented parasites of the respiratory tract of vertebrates; body wall, including a cuticle, is metameric; mite-like larvae and sperm morphology indicate a close relationship to crustaceans; 90 species (*Linguatula*, *Armillifer*).

PHYLUM ONYCHOPHORA

Velvet or walking worms, arthropod-like inhabitants of the moist tropics; unjointed legs and several internal systems are metameric; 80 species (*Peripatus*).

THE ARTHROPODS

The arthropods are, without question, the most diverse and successful animal phylum. Estimates of the number of described species vary, but most figures indicate the presence of approximately 1 million known species and the possibility of between 2 million and 10 million undescribed species. Many of these undescribed species reside in dense tropical rain forests or other areas which are nearly inaccessible to man. Some biologists estimate that the extinction rate due to deforestation exceeds one species per day in the Amazon region alone. The incredible success of these animals stems from a handful of common, ancestral characteristics: 1) a **chitinous exoskeleton,** which covers and protects the body; 2) paired, jointed **appendages** which are often highly modified to perform specific functions; and 3) a **metameric** body in which the individual segments have become fused evolutionarily into functional groups called **tagmata.** These tagmata often form either two or three distinct body regions which specialize in certain functions (feeding and sensory perception; locomotion; digestion and reproduction, etc.) The paired appendages are often the most conspicuous arthropod features; they can be adapted for walking, jumping, swimming, defense, feeding, sensory perception, and even the generation of airborne sounds. The exoskeleton is secreted by the epidermis and must be periodically molted (**ecdysis**) in order for the animal to grow. In addition to imparting structural support, the exoskeleton retards moisture loss and provides an ideal area for muscle attachment.

Other features that characterize the phylum are: the presence of a reduced body cavity (**hemocoel**) which is filled with a blood-like fluid (**hemolymph**); a high degree of cephalization with complex sensory organs, a well-developed brain, and segmented ganglia placed along a ventral nerve cord; a closed arterial system with a dorsal heart, tubular arteries, and sinuses, but an open venous system. Arthropods inhabit nearly every known environment, ranging from deep ocean trenches to the peaks of the Himalayas. While the vast majority of arthropods inhabit temperate and tropical regions, some species have successfully invaded both the arctic and antarctic regions. Arthropods are enormously important to human culture. They greatly influence agricultural crops (through both pollination and destructive feeding), they are important disease vectors (carrying the agents which cause malaria, bubonic plague, lyme disease, typhus, yellow fever, Rocky Mountain spotted fever, and many others), and they make up a large proportion of the energy web in the environments in which they live.

Due to the great diversity and proliferation of species, accepted taxonomic schemes of the Arthropoda vary but most authors recognize four subphyla of arthropods: the Trilobita, the Chelicerata, the Crustacea, and the Uniramia. Trilobites are extinct marine arthropods which dwelled among the sediments of benthic environments. The chelicerates include the terrestrial arachnids (spiders, scorpions, ticks, and mites) as well as the marine horseshoe crabs and the sea spiders. The crustaceans [crabs, shrimp, lobsters, and crayfish (called crawfish or crawdads also!)] are mostly aquatic and are among the most abundant and successful invertebrates. The Uniramia include several highly successful terrestrial groups, notably the insects, the centipedes, and the millipedes.

SUBPHYLUM TRILOBITA

Trilobites are ancient arthropods which once dominated the benthic environments of the ocean floor. Although they have been extinct for 280 million years, many display features typical of living arthropods, such as a chitinous exoskeleton, tagmata, and branched appendages. Trilobites were dorsoventrally flattened and divided longitudinally into three lobes (hence the name of the group) as well as divided horizontally into three tagmata: the anterior **cephalon** (head), the **thorax,** and the posterior **pygidium.** Each body segment bore two biramous (branched) appendages; one branch was adapted for walking/crawling, while the other branch, which was highly filamentous, was possibly used in respiration as a gill. The cephalon usually bears a pair of eyes and sensory antennae. Most trilobites were small creatures averaging less than 10 cm in length; however, some species approached the one meter mark. Observe a fossil specimen of a trilobite and note the characteristic features.

SUBPHYLUM CHELICERATA

The chelicerates are distinguished from the remaining arthropods by the presence of two basic body features: the first pair of appendages are modified feeding organs termed **chelicerae;** the second pair of appendages, the **pedipalps,** function in feeding or sensory perception depending upon the species. Members of the subphylum characteristically lack antennae and the body is divided into two tagmata rather than the more typical three tagmata of insects. The anterior body region, the **prosoma,** bears the chelicerae, the pedipalps, and the walking legs. The posterior **opisthosoma** houses the primary digestive and reproductive organs but lacks external appendages.

Class Merostomata: *Limulus*

Limulus, the horseshoe crab, is a common marine arthropod which inhabits the Atlantic and Gulf coasts of North America (Figure 4.1). The body is typical of a chelicerate, but the dorsal surface is modified to help push the animal through mud and sand in search of its soft-bodied invertebrate prey (polychaetes and molluscs). The animal is protected dorsally by well-developed exoskeletal shields. The anterior shield, the prosoma, is horseshoe-shaped and bears two **compound eyes,** a feature unique to living chelicerates. Anteriorly and medially are two smaller **simple eyes.** The opisthoma possesses protective spines and a long, spine-like telson. Ventrally, *Limulus* bears food handling chelicerae, a pair of pedipalps, and four pair of walking legs. All six paired appendages are **chelate** (pincer-like). The last pair of walking legs is elongate and modified for cleaning the gills and pushing the animal through the sediments.

At the base of the walking legs, located centrally, is the **mouth.** The mouth is surrounded by spinous **gnathobases** which shred the prey item and force food toward the mouth. The ventral surface of the opisthoma houses the **gills** underneath six pairs of broad, flat abdominal plates. The first pair of plates, the **genital opercula,** bears a genital pore through which gametes are shed. Underneath the second through the sixth pair of opercula are highly vascular folds of the **book gills.** These abdominal folds open and close to provide a continuous flow of seawater over the internal gills. Intense beating of the abdominal folds may also provide aid in swimming short distances. *Limulus* is a large arthropod, reaching sizes of up to 60 cm. It is often regarded as a "living fossil" since it is an ancient species which has not significantly modified its morphological features for at least 250 million years.

Class Arachnida:
Argiope and other arachnids

The arachnids comprise a group of 65,000 species of small, highly successful chelicerates. They vary considerably in form, ranging from the diminutive ticks and mites, to the spiders, scorpions, pseudoscorpions, and whip scorpions. Most arachnids are terrestrial species which possess **four pair** of **walking legs, book lungs** (internal respiratory folds modified from book gills), and water-conserving excretory organs, the **coxal glands** and **malpighian tubules.** Arachnids probably evolved from aquatic ancestors and it is these features which allowed them to successfully invade terrestrial habitats. Arachnids are known to occupy nearly every

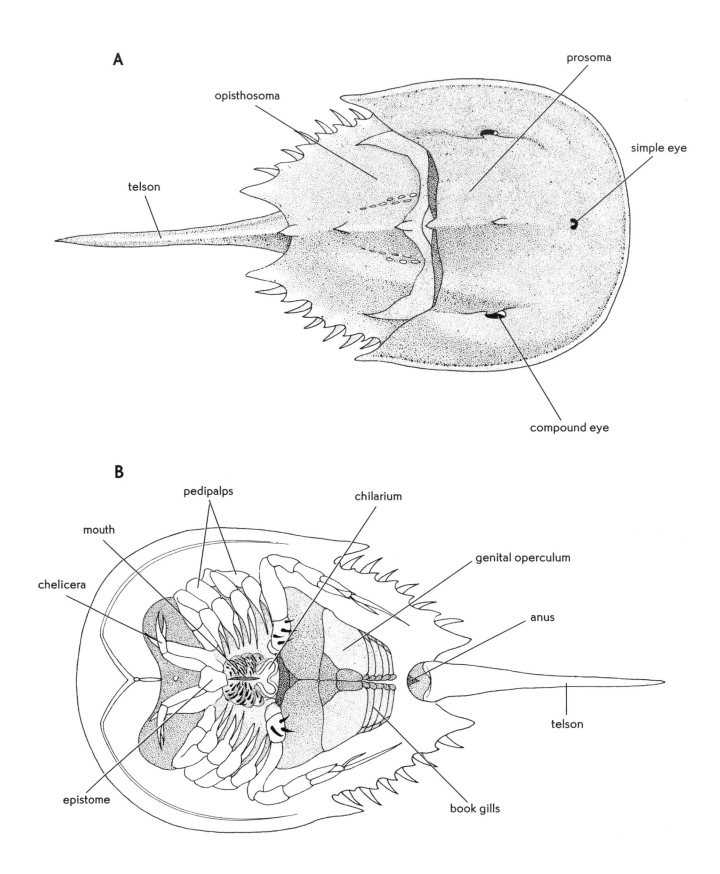

FIGURE 4.1 *Limulus,* the horseshoe crab. A) dorsal surface; B) ventral surface.

available habitat, with the exception of deep marine waters and extreme polar environments.

Argiope

Spiders are carnivores that feed primarily on insects, often devising ingenious ways to capture, trap, or immobilize their prey before ingestion. Many spiders trap their prey with elaborate webs, thus allowing the luxury of feeding at leisure, while other species (jumping spiders, wolf spiders, and trap-door spiders) poison their prey immediately after capture. The chelicerae of most spiders are modified into **fangs** specialized for delivering poison. The pedipalps are leg-like and perform sensory and reproductive functions (the transfer of sperm). Spiders typically have a very slender "waist," the **pedicel,** which separates the prosoma from the opisthosoma. The opisthosoma houses internal **silk glands** which secrete the web through a pair of posterior pores positioned near several small appendages termed the **spinnerets.** Obtain a preserved specimen of *Argiope,* the common garden spider, and locate the preceding features. Many other spiders will display the typical arachnid characteristics while showing some diversity of form. If available, observe several other important spiders such as: the Black widow (*Latrodectus*); the Brown recluse, both of which are highly poisonous; and a *Tarantula,* a large but relatively harmless (to humans) spider common in the southwest United States.

Various Arachnids

Other common arachnids include the **ticks** and **mites, scorpions, pseudoscorpions, whip scorpions,** and **harvestmen.** Harvestmen (daddy longlegs) are often mistaken for spiders due to a superficial resemblance, but harvestmen lack a pedicel, and are omnivorous, eating a mixture of vegetation, fruit, insects, and decaying organic matter. Harvestmen thrive in moist environments and are common in the plant litter on the forest floor.

Ticks and mites are blood-sucking ectoparasites of terrestrial vertebrates. Ticks live on blades of grass and along tree branches waiting for unsuspecting mammals to travel nearby. When a suitable host is near they drop off of the vegetation and cling to the host, forming a stout shaft and blood-sucking tube out of the chelicerae. Teeth on the chelicerae and spines on the pedipalps help the tick grip the host's skin and create a wound. Blood is then sucked into the ticks body over a period of time. Female ticks require a large blood meal to complete egg production, so they often continue feeding until they become quite engorged and swollen. Mites are much smaller than ticks (usually less than 1 mm in length) hence their parasitic microenvironments are more extensive. Mites are known to occupy mammalian hair follicles, reptilian, avian, and mammalian epidermal tissues, internal organs (lungs), insect tracheae, and even the internal auditory canals of vertebrates. Dust mites (*Dermatophagoides*) cause severe allergic reactions in some people, and the notorious "chiggers" (*Trombicula*) are a group of mites in which the larvae inject digestive enzymes into the skin and suck up the fluids from the digested cells. After the chigger (also called "redbugs" or "itch mites") drops off the skin surface a persistant itch may occur which causes the host to scratch incessantly.

Scorpions are nocturnal predators which capture insects and other small invertebrates with their large pedipalps (pincers) which hold the struggling prey until it can be stung by the tail (Figure 4.2). The tail harbors

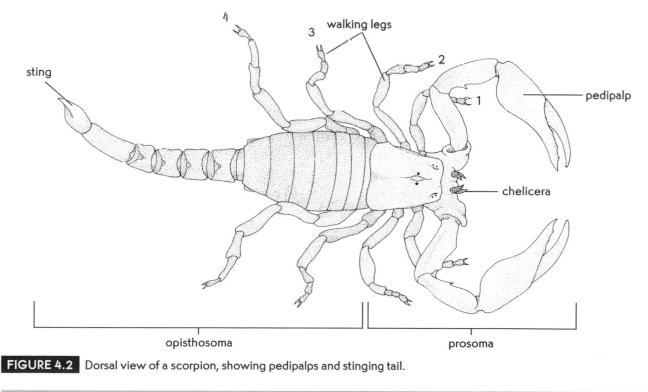

FIGURE 4.2 Dorsal view of a scorpion, showing pedipalps and stinging tail.

EXERCISE 4   *The Arthropods*   **83**

a poison gland and needle-like stinger which injects a powerful neurotoxin into the prey. Most scorpions hide in crevices and burrows during the day and hunt at night. Only a few species are poisonous enough to harm humans. Most of these species encounter humans accidentally, when suddenly uncovered from their hiding places. Pseudoscorpions have large pincer-like pedipalps but lack the stinging tail typical of scorpions. Pseudoscorpions are much smaller than most scorpions, and many species have a dark red anterior coloration. Their pedipalps possess poison glands useful in subduing small insects. Whip scorpions (sometimes called **vinegaroons**) live in warm, humid habitats where they capture small insects and crush them with their large pedipalps. Whip scorpions lack poison glands, but they can secrete a noxious chemical (acetic acid) in defense. The acid, which smells somewhat like vinegar, is released from glands near the long whip-like tail. Whip scorpions average 2–4 mm in length, but the giant American vinegaroon attains lengths of up to 7 cm.

SUBPHYLUM CRUSTACEA

Crustaceans comprise a relatively homogenous group of aquatic arthropods. Of the 40,000 living species, only a few members exhibit a truly terrestrial existence, and even these species must remain moist and are often found hiding in dark, damp areas. The majority of crustaceans are marine, but many ecologically important freshwater species also exist. Small planktonic species and the larvae of many larger species are an important component of aquatic food chains and comprise a large percentage of the zooplankton community. The body plan of crustaceans is relatively uniform, consisting of a **head, thorax,** and **abdomen.** In many species the head and thorax are fused, creating a **cephalothorax.** The appendages are segmentally arranged and usually **biramous** (two-branched). Some of the appendages bear **gills** internally. Crustaceans are covered with an **exoskeleton,** which in large species is often calcified. Two pairs of **antennae** are present, and the head usually possesses a single pair of **compound eyes.** The five pair of head appendages are generally similar throughout the subphylum. The first and second pairs (the **antennules** and **antennae,** respectively) are modified for sensory reception, feeding, or locomotion depending upon the particular species. The third pair of head appendages, the **mandibles,** are opposable mouthparts which usually serve as grinding, chewing, or filtering apparati. Behind the mandibles, lie two additional appendages, the **first** and **second maxillae.** These appendages perform a variety of food sorting, tasting, and handling functions.

The biramous appendages of crustaceans are often greatly modified and show regional specialization. Some species display secondarily derived uniramous appendages, but most follow a basic morphological pattern with modification dependent upon function. The typical biramous appendage consists of a proximal **protopodite** with two distal rami. The inner or medial ramus is termed the **endopodite** and the outer or lateral ramus is called the **exopodite.** The systematics of crustaceans vary considerably depending upon the source, but most authors break the subphylum down into the following five major classes: the **Branchiopoda,** the **Copepoda,** the **Ostracoda,** the **Cirripedia,** and the **Malacostraca.** Each of these classes is ecologically significant, but the malacostracans (shrimp, lobsters, crabs, crawfish) represent nearly 70% of all crustacean species and are the dominant class in both ecologic and economic importance.

Class Branchiopoda:
Artemia, Daphnia and *Eubranchipus*

Branchiopods are small, freshwater crustaceans with segmented bodies. The segments are often obscured by an overlying **carapace** which protects and streamlines the organism. *Artemia,* the brine shrimp, *Daphnia,* the water flea, and *Eubranchipus,* the fairy shrimp, are common branchiopod genera which display features typical of the class (Figure 4.3). Branchiopods usually have a generous supply of appendages which function in respiration (hence the class name which means "gill foot"), filter-feeding, and/or locomotion.

Class Copepoda: *Cyclops*

7,500 species of extremely small (most are less than 1 mm), planktonic crustaceans which are abundant in both marine and freshwater environments. Copepods frequently occur in dense populations, swimming about the water column catching planktonic algae, bacteria, zooplankton, and suspended particles. Nearly 1,500 species of copepod are parasitic, inhabiting the body surface of fish and some marine mammals. Some species inhabit the sediments of benthic environments. *Cyclops,* a freshwater genus, typifies the class structurally (Figure 4.3). The **head** bears a single median **compound eye** and sensory **antennae.** The **thorax** is tapered and possesses several feeding appendages (**maxillipeds**), while the narrow **abdomen** is usually lacking appendages. Similar anatomy is displayed by *Cletocamptus,* a brackish water genus (Figure 4.4).

Class Ostracoda: seed shrimps

Ostracods are enclosed in a bivalved carapace which imparts an appearance similar to that of a bean, mussel, or plant seed. A few species of ostracod are terrestrial, but most species occupy marine or freshwater habitats, filtering organic particles from the water. Their size typically ranges from 1 to 3 mm, but some marine species reach lengths of up to 2 cm. The carapace is strengthened with calcium carbonate ($CaCO_3$; limestone) except in the region of the hinge. The elastic hinge is operated by muscles that close the valves upon contraction, and open the valves upon contraction.

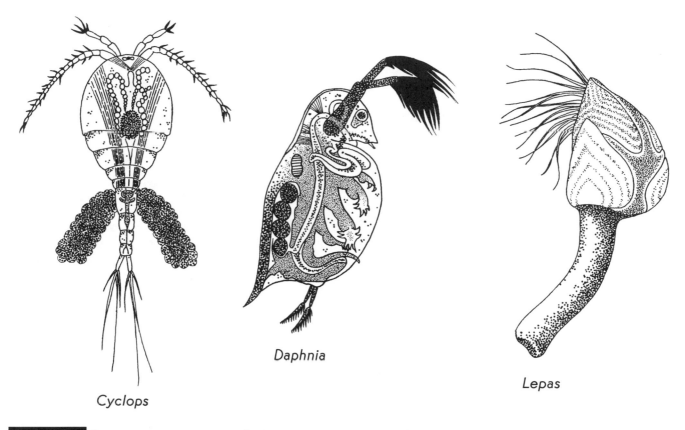

FIGURE 4.3 Representative crustaceans. *Cyclops*, a copepod; *Daphnia*, a branchiopod, and *Lepas*, a cirripedian.

Cyclops

Daphnia

Lepas

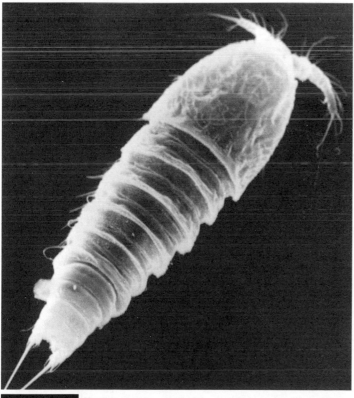

FIGURE 4.4 *Cletocamptus deitersi*, a common brackish water copepod. Scanning electron micrograph by J. W. Fleeger.

Class Cirripedia: *Balanus* and *Lepas*

Cirripedians include nearly 1,000 species of sessile, marine crustaceans which often secrete a heavy calcareous shell. The presence of the shell led early systematists to misclassify cirripedians as molluscs. It was not until 1830, when their larvae were described, that these animals were properly placed within the Crustacea.

The shell is actually a highly modified **exoskeleton.** The cirripedians, commonly called **barnacles,** can be either free-living or parasitic. The free-living forms occur in two basic morphologies: stalked (goose-neck) or stalkless (acorn or volcano) barnacles. Stalked barnacles (*Lepas*) attach to the substrate by means of a flexible, elongate stalk (Figure 4.3). Stalkless barnacles (*Balanus*) attach the shell directly to the substrate and frequently reside in large aggregations. Aggregations can form on any surface in an environmentally suitable area, but most barnacles are found on intertidal rocks and ledges, ship hulls, and wharf pilings. The low-profile, squat appearance of a stalkless barnacle is a modification designed to withstand wave action and strong currents. Compare the structure of *Balanus* with that of *Lepas*. Note the similarities as well as the obvious differences.

Free-living barnacles are filter-feeders that capture nutrients and plankton by waving their appendages (**cirri**) through the water. The cirri can be retracted into the shell and the shell closed to prevent desiccation during periods of air exposure (low tide, etc.).

Class Malacostraca: Isopods, Amphipods, and Decapods (*Procambarus*)

Malacostracans all possess several (one to three) pairs of thoracic appendages termed **maxillipeds**. The maxillipeds are modified for a variety of feeding purposes, including handling, tasting, and sorting of food items. Posterior to the maxillipeds are the thoracic **walking legs**. In many species, the first pair of walking legs are enlarged and form grasping claws termed **chelipeds**. The distal tip of the chelipeds bears terminal pincers, termed **chelae**. The remaining thoracic legs are adapted for crawling along the substrate and give rise to internal **gills**. Of the many diverse malacostracan groups, the three major orders are covered: the Isopoda, the Amphipoda, and the Decapoda.

Isopods

Isopods inhabit a variety of environments. They are most common in marine habitats, but some species are found in freshwater, semiterrestrial, and truly terrestrial environments. Isopods possess a single pair of maxillipeds, are often flattened dorsoventrally, and lack a carapace, a trait which exposes the individual segments. Their respiratory surfaces lie beneath the exoskeleton, and are attached to the abdominal appendages. Respiration can be accomplished as long as the animal remains moist, thus, some isopods represent the most successful terrestrial crustacean species. The pill or sow bug (*Porcellio, Oniscus*) is a common terrestrial isopod found in damp areas under rocks or decaying wood. Here it feeds on decaying organic matter and some living plant and animal material. The giant marine isopod (*Bathynomus giganticus*) is a relative of the pill bug which dwarfs its terrestrial relatives, reaching lengths of 30–40 cm.

Amphipods

Similar to isopods in appearance and ecologic factors, amphipods possess several isopod-like features. They are primarily marine, but freshwater and semiterrestrial species also exist. They possess only one pair of maxillipeds, and they lack a carapace. Unlike isopods, however, amphipods are flattened laterally rather than dorsoventrally, and they possess thoracic rather than abdominal gills. Dense populations of freshwater amphipods are common in streams and ponds. Beach environments provide an ideal habitat for semiterrestrial amphipods like beach fleas and sand hoppers.

Decapods

Comprised of 10,000 species of crabs, shrimp, lobsters, and crawfish, this is the largest order of the Crustacea.

Many species are marine, but several highly conspicuous species (such as crawfish and some shrimp) are freshwater, and a few species (burrowing crabs and crawfish) even live on land. Crawfish, lobsters, and crabs are generally adapted for walking and crawling across a surface, whereas shrimp are adapted for swimming through open water. Decapods, as the name indicates, possess ten walking legs (five pair). The first pair are often chelate. The head and thorax are fused into a single **cephalothorax** which is covered with a carapace enclosing lateral gill chambers.

Crabs generally follow a broad and flat body plan, with a reduced abdomen flexed under the cephalothorax. This allows the animal to move more efficiently over a surface because it does not have to drag its abdomen as lobsters and crawfish do. Crabs are often abundant in intertidal zones and crowd beach environments. Many species are relatively small, but some crabs, like the Alaskan king crab, can reach lengths approaching 4 meters. Observe several crab specimens on demonstration. *Pagarus*, the hermit crab, is a common species which resides in empty gastropod shells, carrying them around on their backs for immediate protection from predators. *Mennipe*, the stone crab, possesses a large cheliped which it uses to break open mollusc shells during predation. Watch out, this species can break your finger if accidentally caught in the cheliped! *Libinia*, the spider crab, is a bizzare-looking crab which superficially resembles an arachnid with its spines and protruding legs. *Callinectes*, the blue crab, is an important commercial species, providing food for many coastal inhabitants. Male fiddler crabs, genus *Uca*, have a single enlarged cheliped which the male uses to lure a female into his burrow.

Shrimp are important from an economic and ecologic standpoint. Many marine species (*Penaeus*) are relished as a food source, and freshwater species (such as grass shrimp) are often sold as fish bait. Lobsters live among rocks and coral reefs on the seafloor where they scavenge for clams, snails, and crabs. The American lobster, *Homarus americanus*, is one of the largest living arthropods reaching lengths of 50 cm and weighing in at 20 kg.

 **PRESERVED SPECIMENS**

Procambarus

Crawfish are freshwater relatives of the lobster, but do not quite reach the immense size of most lobsters. Many species of crawfish, such as *Cambarus* and *Astacus*, average 5–10 cm, but the southern crawfish, *Procambarus clarkii*, reaches lengths of 15 cm and is a common food item in Louisiana. Most crawfish (300 species worldwide, 100 in the U.S.) are omnivores, feeding on a variety of soft-bodied organisms such as worms, snails, fish, small frogs, and tadpoles. Plant matter and decaying organic material are often favorite foods, and live crawfish can be seen sorting, tasting, and handling minute food particles almost constantly.

External anatomy. Place a crawfish in a dissecting pan with the dorsal surface up. The **exoskeleton** is a cuticle secreted by the epidermis and hardened with a nitrogenous polysaccharide (**chitin**), and the addition of **calcium carbonate** (limestone). The cuticle must be shed or **molted** (**ecdysis**) several times while the crawfish are maturing. Each time the exoskeleton is discarded, the animal must secrete new materials in order to complete growth. The new soft exoskeleton soon hardens. The exoskeleton is differentiated into two tagmata, a fused **cephalothorax** and a segmented **abdomen** (Figure 4.5).

A hard **carapace** covers the cephalothorax. It is delineated by grooves into four regions: the **head**; a dorsal, median **cardiac region** lying over the heart; and two lateral branchial regions that cover the gills. Lift the lower edge of the carapace to reveal the **gill chamber** with the fimbriate **gills.** Each pair of gills is attached to the base of the walking legs which extend laterally from the cephalothorax.

At the anterior tip of the head a pointed **rostrum** lies between a pair of stalked, movable **compound eyes.** The small surface **facets** of the eye are terminations of the visual cells of the eye. The short **antennules** and the long **antennae** bear tactile (touch) and chemosensory organs.

The head bears five pairs of appendages: the antennules, the antennae, and three pairs of mouthparts (the **maxillae** and **mandibles**). The thorax bears three pairs of **maxillipeds** and five pairs of long **walking legs.** The abdomen is made up of six segments. Each is covered with a hard dorsal **sternum.** The abdomen terminates with a flat median process, the **telson,** bearing the **anus** on the ventral side. On either side of the telson, locate the lateral **uropods.** Both the telson and uropods aid in

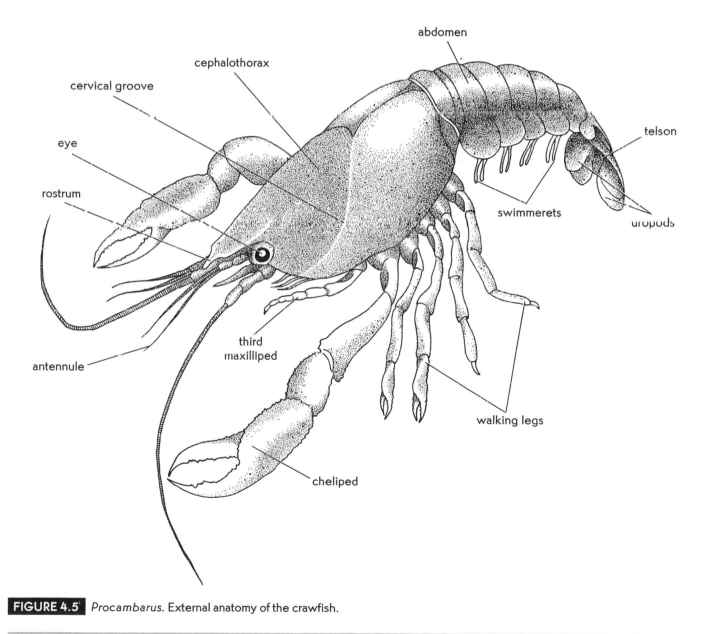

FIGURE 4.5 *Procambarus.* External anatomy of the crawfish.

swimming and are designed to quickly curl under the abdomen when the animal is reversing directions or protecting the eggs and/or larvae.

The first five abdominal segments bear **swimmerets (pleopods)**. Each swimmeret is biramous and sexually differentiated. In the male, the first two pairs of swimmerets are modified into stiff, grooved structures used to transfer sperm to the female. In the female, the first pair is reduced in size and the remaining pairs are soft, fleshy, and well-developed. Here the female will brood her eggs, attaching them to the swimmerets for protection and increased oxygenation. The **genital openings** also show sexual dimorphism. In the male, the genital openings of the sperm ducts are located medially at the base of each of the fifth pair of walking legs. In the female, the genital openings of the oviducts are located at the base of each of the third pair of walking legs. Also, the female possesses a midventral opening to the seminal receptacle. During copulation the male turns the female over and sheds the sperm directly onto her abdomen.

The **appendages** of crawfish are highly modified and well-developed. Remove and study each of the appendages shown in Figure 4.6. To remove an appendage, grasp it firmly at the base with a pair of forceps and gradually work it loose from the body. The appendages are numbered consecutively beginning at the anterior end of the animal. The **antennules**, which function in maintaining equilibrium, have a three-jointed protopod and two long highly-jointed filaments. The **antennae** are sensitive to touch, vibrations, and chemical concentrations in the water. The endopod of each antenna is a long, many jointed filament and the exopod is a broad, movable projection near the base. The protopod bears the **renal opening** which expels waste collected by the internal antennal (green) gland. The **mandibles** are located on either side of the mouth and are heavy, grinding structures which possess teeth along the inner margin. This represents the protopod; the exopod is absent, and the endopod is a small palp folded above each mandibular "tooth." The mandibles grind from side to side, holding the food while the maxillipeds shred it. Pull the mandible out and note the muscle attachment.

The **first** and **second maxillae** are small head appendages, located just external to the mandibles, that place food in the mouth. The first maxillae is reduced and foliaceous; the second maxillae forces water currents toward the gill chamber with its expanded exopod, the **gill bailer**. The three pairs of **maxillipeds** all possess a large endopod and a slender, filamentous exopod. The **first** and **second maxillipeds** are much smaller than the **third maxilliped**. All are food handlers which channel and direct food to the mouth, but the first maxilliped bears no gill.

The **walking legs (pereiopods)** are specialized for locomotion, defense, and food handling. All of the walking legs are uniramous and lack the exopod. The first pair of legs possess enlarged claws and are termed **chelipeds**. The distal pincers, used in territorial battles and defense, are called **chelae**. The remaining four pairs are leg-like and used in walking and food handling. The first four pairs bear gills internally; the second and third pairs possess small chelae used for grasping. The **swimmerets (pleopods)** are biramous organs that function in swimming, reproduction, and the creation of water currents that pass over the gills. The single proximal portion is the protopod, the distal branched portion consists of the medial endopod and lateral exopod. The medial **telson** is surrounded by the lateral **uropods**, which, again, are biramous.

Internal anatomy. To dissect the crawfish, you will first need to carefully remove the carapace. This can be accomplished by inserting the point of the scissors under the posterior edge of the carapace about near the medial line. Cut forward, staying as close to the exoskeleton as possible, until the rostrum has been cleanly split. Carefully pull down on the anterior-ventral edge, removing the carapace from the rest of the body. Be sure to notice the underlying **epidermis** and **muscles**, which attach to the carapace, especially in the head region (Figure 4.7). Secure any loose epidermal tissue with your scalpel and push it carefully back into place. Remove the dorsal portion of the abdominal exoskeleton in the same way, uncovering each somite carefully so as not to destroy the long **extensor muscle** lying underneath.

Circulatory system. The thin tissue covering the viscera is the **epidermis**, which secretes the materials deposited in the exoskeleton. Note the location of the pigmented (often red or pinkish) portion of the epidermis; this indicates the position of the **cardiac region** of the carapace. This tissue covers the **pericardial sinus** which surrounds the heart. If the specimen has been injected with latex (to facilitate study of the circulatory system), the sinus may be filled with a red rubbery substance. The small, angular **heart** is located dorsally, just anterior to the junction between the cephalothorax and the abdomen. If the specimen has been injected, the heart may lose its distinctive shape and the **ostia** (openings in the heart) may be obscured. Carefully grasp the heart with a pair of forceps and remove it from the sinus. Note the small vessels which are associated with the heart. Crawfish have an **open circulatory system**. The **hemolymph** (blood) leaves the heart in arteries but returns to it by way of venous sinuses, or cavities, rather than enclosed in veins. Hemolymph enters the heart through three pairs of slit-like openings, the ostia, which open to receive the hemolymph and then close when the heart contracts to force the hemolymph out through the arteries.

The arterial system can be seen in a quality dissection, but the availability of a model may enhance observation. Five arteries leave the anterior end of the heart: a median **ophthalmic artery** extends forward to

APPENDAGE   HEAD and THORAX   FUNCTION

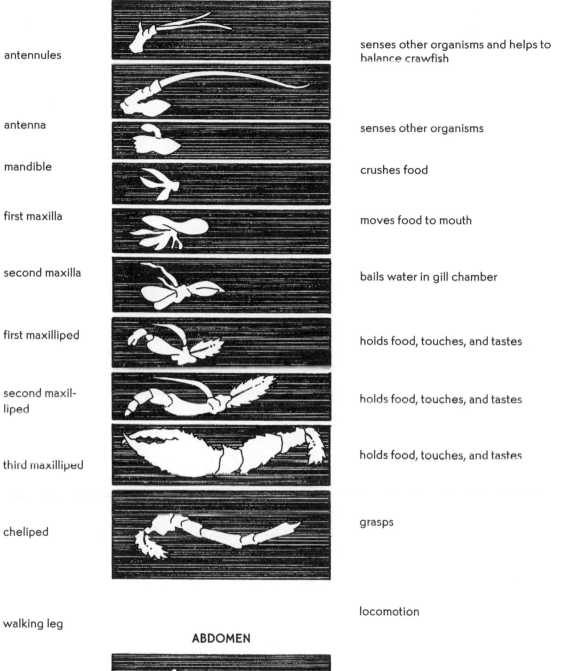

antennules   senses other organisms and helps to balance crawfish

antenna   senses other organisms

mandible   crushes food

first maxilla   moves food to mouth

second maxilla   bails water in gill chamber

first maxilliped   holds food, touches, and tastes

second maxil-liped   holds food, touches, and tastes

third maxilliped   holds food, touches, and tastes

cheliped   grasps

                locomotion

walking leg

ABDOMEN

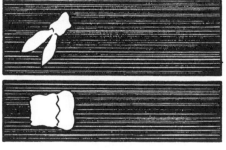

first swimmeret in male transfers sperm to female, which uses 2nd, 3rd, 4th, and 5th swimmerets to hold eggs

FIGURE 4.6 Serial appendages of the crawfish, *Procambarus*.

supply the cardiac stomach, the esophagus, and the head; a pair of **antennal arteries,** one on each side of the ophthalmic, pass diagonally forward and downward over the digestive gland (hepatopancreas) to supply the stomach, the antennae, the antennal glands, and parts of the head; and a pair of **hepatic arteries** from the ventral surface of the heart supply the hepatopancreas.

Exiting the posterior end of the heart are the **dorsal abdominal artery,** which extends the length of the abdomen and lies on the dorsal side of the intestine, and the **sternal artery,** which runs ventrally to the nerve cord where it divides into the **ventral thoracic** and **ventral abdominal arteries,** which supply the appendages and other ventral structures.

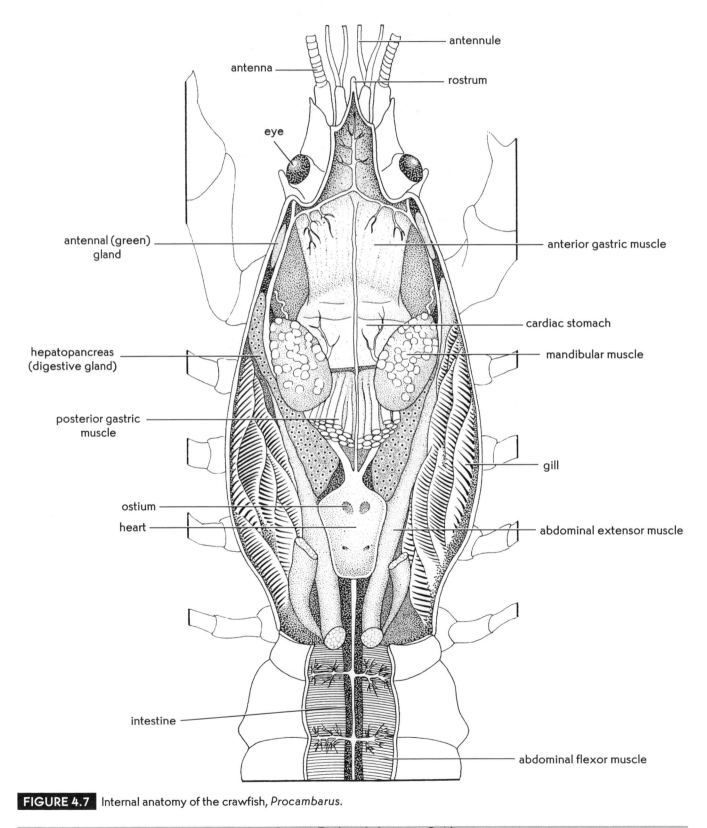

FIGURE 4.7 Internal anatomy of the crawfish, *Procambarus*.

Digestive system. If the epidermis is obscuring the viscera, very carefully pull it off with a pair of forceps. The large **stomach** lies in the head region, just behind the rostrum, anterior to the heart. It is large, thin-walled, and usually dark colored due to the presence of sediment and organic matter obtained during feeding. The stomach consists of two portions; the anterior **cardiac chamber** and a smaller, softer posterior **pyloric chamber.** With a pair of forceps, pull on the stomach dorsally to remove it from the body. You can now see the **esophagus,** a short tube connecting the cardiac chamber to the ventral, external mouth. Cut the stomach open and rinse out the digestive material. Note the presence of two sets of **gastric teeth** and a single, median grinding mill. These structures function in mechanical digestion and facilitate preparation of food for absorbtion. You may find a mass of calcareous crystals (**gastroliths**) attached to the sides of the cardiac chamber. These hard conglomerations have been extracted from the old exoskeleton by the blood and are source matrerial used in the formation of the new exoskeleton. These are only visible during molting periods.

On either side of the stomach and heart are the large yellow or cream-colored lobes of the **hepatopancreas (digestive gland).** The distal lobes extend to the posterior region of the thorax, and as typical of many animals, this is the largest organ in the body. The hepatopancreas produces digestive secretions that are dumped into the pyloric chamber of the stomach through hepatic ducts. The hepatopancreas is also an important site for absorption and provides storage of food reserves.

The **intestine** is a small, dark tube which leaves the pyloric chamber, passes under the heart, and then runs dorsally and posteriorly along the abdomen between the flexor muscles. In the sixth abdominal segment it expands to form a dead-end sac, the **cecum,** and terminates at the **anus.** The anus is located externally on the ventral surface of the telson.

Reproductive system. In both sexes, the **gonads** are small, inconspicuous organs located just underneath the heart. Their size and prominence will depend largely on the season in which the animals were captured before preservation. The gonads are very slender organs usually slightly different in color and texture from the digestive gland, lying along the medial line between and the distal lobes of the gland. The gonads are often difficult to distinguish from the digestive glands, and may not be readily visible during nonbreeding periods. The male gonads (**testes**) are delicate organs with a whitish appearance. The **sperm ducts** pass diagonally over the digestive glands and back to the gential openings associated with the fifth pair of walking legs. In the female, the **ovaries** are slender and orange colored. In the reproductive season the ovaries become swollen and greatly distended with eggs. A pair of **oviducts** passes laterally over the digestive glands to the genital openings at the base of the third pair of walking legs.

Muscular system. When you first removed the carapace, you may have seen two large red bundles of muscle on either side of the stomach. These are the **mandibular muscles** which open and close the mandibles. On each side of the cephalothorax, just under the carapace, a narrow band of muscle runs longitudinally to the distal end of the abdomen. These are the **abdominal extensor muscles** that, upon contraction, straighten the abdomen region. In the abdomen, lying beneath the extensors and nearly filling the abdomen, are the **abdominal flexor muscles,** which bend the abdomen ventrally, and cause the abdomen to curl. The flexors are large and important muscles used in the quick backward escape movements commonly displayed by the animal.

Excretory System. In the head region, anterior and ventral to the digestive glands and lying on either side of the mouth, are a pair of excretory organs. These **antennal glands** (or **green glands,** will not appear green in the preserved material), are round and disc-shaped. The glands each contain an **end sac,** connected by an excretory tubule to a bladder. Coelomic fluid is filtered into the end sac by hydrostatic pressure in the hemocoel. As the filtrate passes through the excretory tubule, reabsorption of salts and water occurs, leaving the urine to be excreted. A duct from the bladder empties through a renal pore at the base of the antenna.

Because crawfish are freshwater animals and water conservation is not an important physiological function, urine is abundantly produced and hypotonic. This also facilitates the storage of salts usually excreted by the urine. In the marine lobsters, which live in an isotonic medium and do not experience salt conservation problems, the nephridial tubule is missing, and the urine is isotonic and produced in low volumes. The role of the antennal glands centers primarily on osmoregulation. Excretion of nitrogenous wastes (mainly ammonia) occurs by diffusion in the gills and across thin areas of the cuticle.

Nervous system. After you have carefully removed the viscera, locate the major structures comprising the nervous system. The brain consists of a pair of **supraesophageal ganglia** that lie against the anterior body wall between the antennal glands. Nerves extend from the ganglia and run to the antennae, antennules, and the eyes. Two connective nerves pass around the esophagus, one on each side, and unite at the **subesophageal ganglion** on the floor of the cephalothorax. Calcified plates that cover and conceal the double ventral nerve cord in the thorax can be removed by careful dissection. Remove the abdominal flexor muscles in the abdomen and trace the **ventral nerve cord** through the length of the body. Locate the ganglia, which appear as enlargements of the cord at intervals which correspond with the abdominal segments, and the small lateral nerves which extend from the ganglia.

SUBPHYLUM UNIRAMIA

Uniramians are so-named based on the presence of unbranched (uniramous) appendages, as opposed to the biramous appendages of crustaceans. Uniramians are primarily terrestrial, but a number of secondarily derived aquatic species exist. Insects make up the vast majority of species, but this subphylum also includes the centipedes and millipedes (collectively called the myriapods) also. The head appendages, which are diagnostic to the subphylum, consist of a single pair of **antennae,** one pair of **mandibles,** and one or two pairs of **maxillae.**

Class Diplopoda: millipedes

Millipedes are omnivorous animals which are capable of digesting a variety of decaying organic matter (Figure 4.8). They are common inhabitants of the forest floor where they feed on the leaf litter. With a cosmopolitan distribution, millipedes play an important ecological role by helping to decompose detritus. Millipedes are slow-moving and nonaggressive but harbor several defense mechanisms. When disturbed, they often coil into a tight spiral for protection and many species produce a noxious chemical which can be sprayed up to 1 meter away. Birds are common predators of millipedes, and many birds have become blind as a result of being sprayed by an upset millipede. Millipedes are highly segmented, cylindrical animals which can be easily distinguished from their relatives, the centipedes, based on a few characteristic features: millipedes possess two pair of legs per body segment and have a single pair of maxillae.

Class Chilopoda: centipedes

Centipedes are flattened, swift-moving, active predators which feed on a variety of small invertebrates (Figure 4.8). Large tropical species, which may approach sizes nearing 1 meter, are known to capture small frogs, toads, and even lizards. The first pair of appendages, the maxillipeds, function as poison claws and immobilize

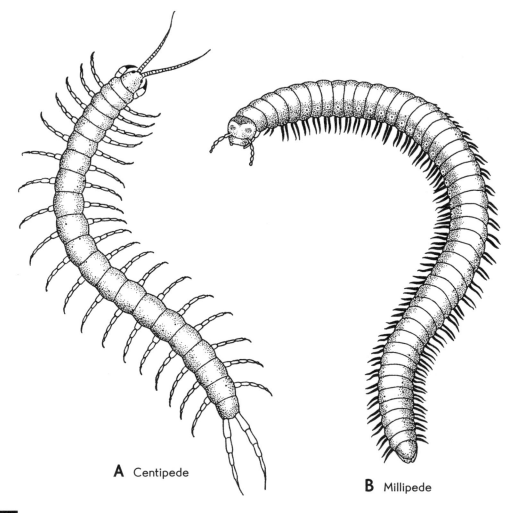

A Centipede

B Millipede

FIGURE 4.8 Myriapods. A) Class Chilopoda, a centipede; B) Class Diplopoda, a millipede.

prey items before the mandibles grind up the food for ingestion. Centipedes possess a single pair of antennae, a single pair of legs per segment, and two pairs of maxillae. Although poisonous, most centipedes are harmless to humans. The house centipede, *Scutigera*, is a common resident of dark, damp household areas, particularly basements, where it helps control insects.

Class Insecta:
Acheta, Apis, and insect orders

Forming the most successful and ecologically dominant group of living organisms, the insects are united by a handful of common features: three pair of legs; a single pair of antennae; generally two pair of wings; and the body divided into three tagmata, the **head, thorax,** and **abdomen.** The head segments have been modified into a protective capsule which bears feeding and sensory structures. Both **compound eyes** and several simple eyes, or **ocelli,** are present. The thorax is the locomotory center of insects, bearing all the legs and wings. The abdomen typically houses the majority of the internal organs, including the digestive, reproductive, and excretory organs.

Insects lack a complex circulatory system and hence must nourish the organs through diffusion of materials in the hemolymph. The respiratory system of insects is unique among the animal kingdom. Oxygen is delivered directly to the organ systems via **tracheal tubes** (Figure 4.9). These tubes comprise a canal system that originates as a series of external abdominal openings termed **spiracles.** The tubes ramify to individual organs and gas exchange occurs through diffusion. This respiratory mechanism is a primary limiting factor of insect size; insects cannot greatly increase their size evolutionarily because they are constrained by the slow diffusion rate of gases through the trachea.

The incredible success of insects can be linked to a unique combination of factors, including their small size, ability to fly, protective exoskeleton, reproductive fecundity, and a metamorphic style of development. With a few notable exceptions, most insects are less than 6 cm in length. This small body size allows insects to occupy habitats that are inaccessible to larger organisms, to feed on food sources too small to sustain bigger animals, and facilitates travel on wind currents and the surface of many larger host organisms. Flying ability enhances the efficiency of predator avoidance, food gathering, and mate selection, as well as allowing the species to disperse across great expanses of habitat. The exoskeleton not only provides protection for the insect from impact and desiccation, but also can be greatly modified to form cutting, piercing, sucking, or grinding mouthparts. Many insects have hair-like **setae** which project from the exoskeleton and aid in locomotion or protection (Figure 4.10).

A short life cycle, the ability to form large numbers of eggs quickly, and rapid growth are all factors which allow insects to reproduce at a fast rate. Some species can cycle through 20–25 generations per year, others

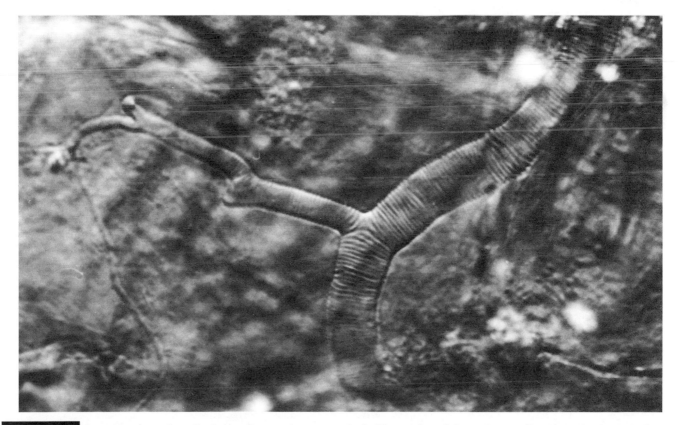

FIGURE 4.9 Insect tracheae from the louse, *Geomydoecus centralis*. These tubes deliver oxygen directly to the organ systems.

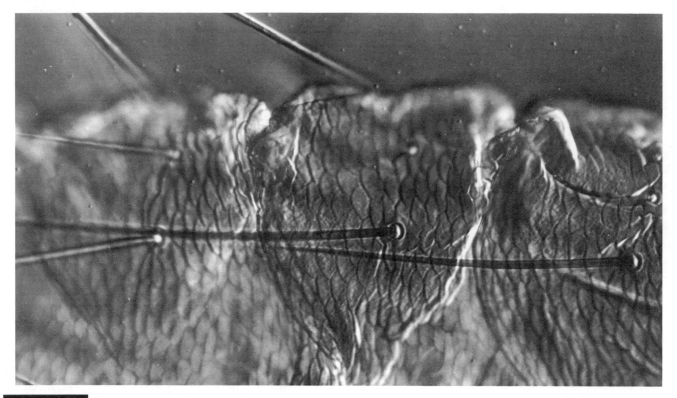

FIGURE 4.10 The integument of the louse, *Geomydoecus centralis*, showing epidermal plates and setae. The length and placement of the setae are important taxonomic tools in this genus.

can lay hundreds of eggs per day during the reproductive season. Lastly, the offspring of adult insects do not compete with the parental generation for resources. Insects often have larval forms which are extremely different in morphological, physiological, and behavioral characteristics, thus the offspring often feed on alternate food sources and may occupy a different habitat than the adults.

The class is divided into 25–30 orders (depending upon the source) which are differentiated based on differences in wing and mouthpart structure. Nine of the major orders will be briefly discussed and representatives of these will be on display. You are probably already familiar with many common insects which typify these 9 orders. (See next page.)

Acheta

The common house cricket, *Acheta domesticus*, is a familiar orthopteran which is native to Europe (Figure 4.11). Sold in the U.S. as fishing bait, fish, amphibian, and reptile food, and also used as a research animal, the house cricket and the sounds they produce have become commonly known. Although the adults have functional wings, the flight muscles degenerate, and crickets typically do not fly. The adult males sing, or "chirp," to attract mates by rubbing portions of their wings together. Adult crickets live an average of 2 months, during which time the female will deposit up to 200 eggs which hatch in 13 days. At 30°C (the ideal incubatory temperature) there are 8 larval stages

(**instars**) before the final molt to the adult stage 48 days later. Crickets cannot climb smooth surfaces (like glass or a strip of aluminum foil) so house crickets make an ideal insect to facilitate live study of both external and internal anatomy.

EXTERNAL ANATOMY

Sex determination and developmental stage. Male and female crickets are easily distinguished based on the presence or absence of the female **ovipositor** which extends from the posterior end of the animal (Figure 4.11). The ovipositor is an elongate rod which is used during egg laying to create cavities in the ground and guide the eggs into the cavity. Both males and females bear soft, flexible posterior appendages termed **cerci.** The developmental stage of the insect can be determined based upon the wing structure. Only adult insects have fully developed, functional wings. In crickets, males have a slightly different wing shape than females. In order to grow, insects must periodically grow a new cuticle and molt the old one. After the egg hatches the insect is a **first instar larva** and after the first molt it becomes a **second instar larva.** In insects, the last two to three larval instars of superorder Exopterygota have external wing pads (buds), such as seen in the cricket larvae. In the superorder Endopterygota the last two to three larval instars have internal wing buds, such as seen in caterpillars, grubs, and maggots. In the exopterygota, the last larval instar molts to the adult; in the endopterygota the last larval instar molts to a subadult

INSECT ORDERS

Order Thysanura (silverfish)

Silverfish are a group of primitive, wingless insects which lack metamorphic development. They are typically small (less than 1 cm) and prefer dark, damp areas. They are common household pests, frequenting basements and bathrooms.

Order Odonata (dragonflies and damselflies)

Odonates possess long, transparent, membranous wings which are supported by a network of veins. Both pair of wings are similar in size and structure. The abdomen is thin and elongate. The larval forms are carnivorous animals common in aquatic habitats where they prey on a variety of invertebrates, small fish, and even tadpoles. Adults are aerial organisms, capturing other insects during flight. Damselflies resemble dragonflies, but possess a thinner abdomen and can fold their wings when resting.

Order Hemiptera (true bugs)

The "true bugs" all possess piercing-sucking mouthparts which are used to suck up blood and tissue fluids from plants and animals. The hindwings and distal half of the forewings are membranous, but the proximal half of the forewings are leathery and protective. The name of the order means "half wing," referring to the appearance of the forewings. The wings generally lie flat over the dorsal surface. This group includes the stinkbugs, waterbugs, water striders, and bedbugs among many others.

Order Homoptera (cicadas, leafhoppers, and aphids)

Homopterans are primarily plant parasites which use piercing-sucking mouthparts to suck the fluids out of plant cells. Dense populations can cause severe stress to even a large plant and many species transmit disease through their saliva. The wings are held at an angle to the body, in a roof-like configuration.

Order Coleoptera (beetles)

The beetles form a large and diverse order comprising 300,000 species, all of which possess a protective sheath (an **elytra**) formed by the forewings. Their hindwings are membranous and used for flying. Most beetles are carnivorous or herbivorous, but a few scavenge food and some species survive by eating fungi. Their habitats are extremely varied; beetles are found in most terrestrial environments and many freshwater streams and ponds.

Order Lepidoptera (butterflies and moths)

Lepidopterans are characterized by having two pair of well-developed membranous flying wings. The wings are formed by a series of tiny overlapping scales (lepidoptera means "scale wing"); usually the hindwings are slightly smaller than the forewings. The larvae of this order are herbivorous, wingless caterpillars which crawl about feeding upon leaves, stems, and fruits. Near the end of the larval stage, most caterpillars weave silk **cocoons** which protect the animal until metamorphosis. The adults bear tubular mouthparts which are modified to suck up nectar from flowers and can be coiled when not in use.

Order Diptera (flies and mosquitoes)

Dipterans are generally excellent flyers despite possessing only a single pair of wings. The hindwings have become reduced and function as a balancing organ, termed a **haltere,** that aids in maneuverability. The larval forms (maggots and wigglers) reside in a variety of habitats, including aquatic environments, inside plant tissue, or in decaying animal carcasses. The adults, with their piercing-sucking mouthparts, feed on blood, tissue fluids, nectar, and other insects. Many are important vectors of disease transmission.

Order Orthoptera (grasshoppers, crickets, walking sticks, etc.)

A large and familiar group of insects, the orthopterans have an inner pair of membranous flying wings protected by an outer pair of leathery forewings. The wings can be reduced or absent in some species. Most species are herbivorous, but omnivory and carnivory are also known.

Order Hymenoptera (ants, bees, and wasps).

Hymenopterans are highly specialized insects which show a great degree of complex social behavior. Many species are important plant pollinators and some are parasites of other insects. The body bears two pair of membranous wings which are hooked together to increase flying efficiency. The thorax and abdomen are separated by a narrow waist, and two large compound eyes are present.

---

called the **pupa.** The pupa then molts (called **eclosion**) to the adult. Adult insects never molt again.

External morphology and dissection. Pin an adult cricket, lateral side up, in a dissecting pan small enough to fit under a dissection microscope. The dorsal surface is called the **tergum,** the sides of the body make up the **pleuron,** and the ventral surface is the **sternum.** Note the three body regions: the **head** (5 segments), the **thorax** (3 segments), and the **abdomen** (10 segments). Feeding and sensory organs are concentrated on the head, locomotory organs (wings and legs) extend from the thorax, and the digestive and reproductive organs are housed in the abdomen. The head segments are fused to form a unit head capsule, and the mouthparts represent the segmentally paired appendages. Only the mesothoracic and metathoracic segments bear wings. The abdominal segments are numbered 1 through 9 (the cerci represent segment 10).

Legs, wings, and mouthparts. Examine the three pairs of legs and identify the **prothoracic, mesothoracic, and metathoracic legs** (Figure 4.11). Cut off a prothoracic leg and identify all segments (**femur, tibia, tarsus**) and the **tympanic membrane** (Figure 4.12). There are two pair of wings, the inner **mesothoracic** and outer

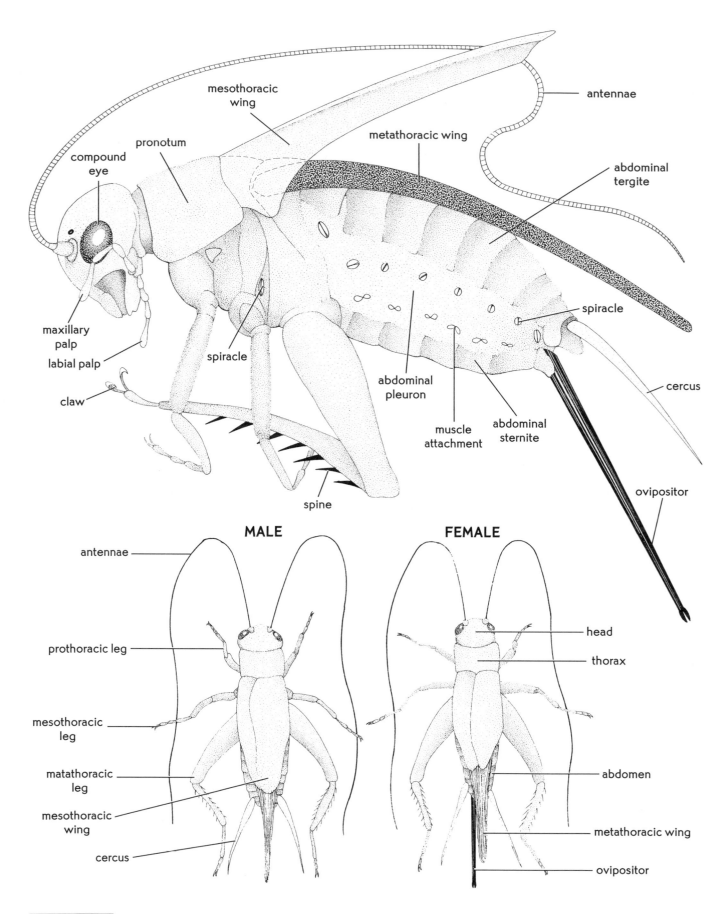

FIGURE 4.11 External anatomy of *Acheta domesticus,* the house cricket.

metathoracic wings. Cut off the mesothoracic wing of a male cricket to observe the **file** and vibrating regions used in the creation of sound (Figure 4.12). The male elevates both forewings up and crosses them, then uses a sawing motion to move one over the other to produce the chirping sound. The rate of chirping is temperature dependent and can be used to estimate the ambient temperature (°F) by counting the number of chirps in 14 seconds and multiplying the product by 4. Place a cricket ventral side up, bend the head back, stick a pin through the inside of the **labrum** (upper lip) to force the head back (Figure 4.12). Identify all mouthparts by moving each with a pair of forceps. By removing a **maxilla** and the **labium** (bottom lip) you will see the **hypopharynx,** which functions as a tongue and bears the openings of the salivary glands.

Feeding, breathing rate, and heartbeat rate. Expose a live cricket to about 5–10 seconds of carbon dioxide (CO_2) to anesthetize it. Dry ice vapors or a CO_2 cylinder can be used as a source of CO_2; these will be provided by your lab instructor. Cut off the wings and immobilize

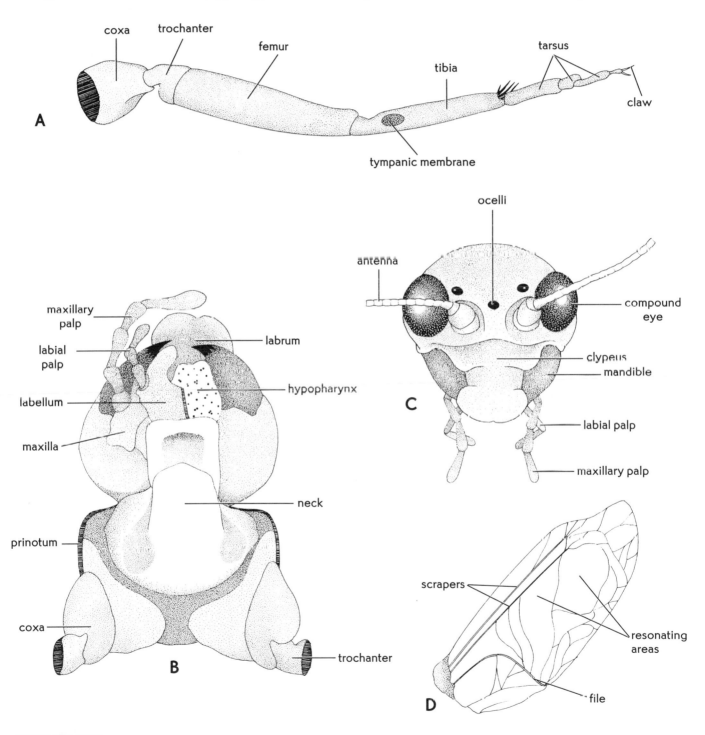

FIGURE 4.12 *Acheta* anatomy. A) prothoracic leg; B) head, in ventral view; C) head, in frontal view; and D) wing structure.

the cricket dorsal side up by using crossing pins over the insect (do not stick the pins through the cricket). When the cricket recovers from the anesthetic, determine the respiratory rate (about 10–20 abdominal contractions per min) and the heartbeat rate (90–120 contractions per min). The heart is a thin, almost transparent tube visible through the intersegmental membranes along the middorsal region. Reanesthetize the cricket, and cross-pin it without injury **ventral side up.** When it recovers, feed it some moistened "Cricket Chow" (Purina) or other suitable food and observe the action of the mouthparts under the dissecting microscope.

General internal anatomy and dissection. Reanesthetize (30–60 seconds of CO_2) a cricket and cut off the wing. Pin it through the head and **epiproct** dorsal side up onto the wax in the dish. Flood the entire animal with

cricket Ringers (a salt solution that matches cricket blood, see below for recipe), and cut the cricket from epiproct to the neck with a pair of fine scissors. Spread open the body wall and pin in place. Be sure the internal organs are covered with Ringers to keep them alive. Observe the peristalsis of the gut and the movement of the malpighian tubules (excretory organs). The tissues will remain viable under these conditions for several hours. The chalky white fat body (insect liver) is spread throughout the body and varies with age and diet of the cricket. Females full of eggs have converted all of the stored food in the fat body to eggs and will have very little fat body. Compare them to a male or larva. The silvery tubes (due to contained air) are the **tracheae,** which ramify to all parts of the body to supply oxygen directly to the tissues. Identify the internal structures presented in Figure 4.13.

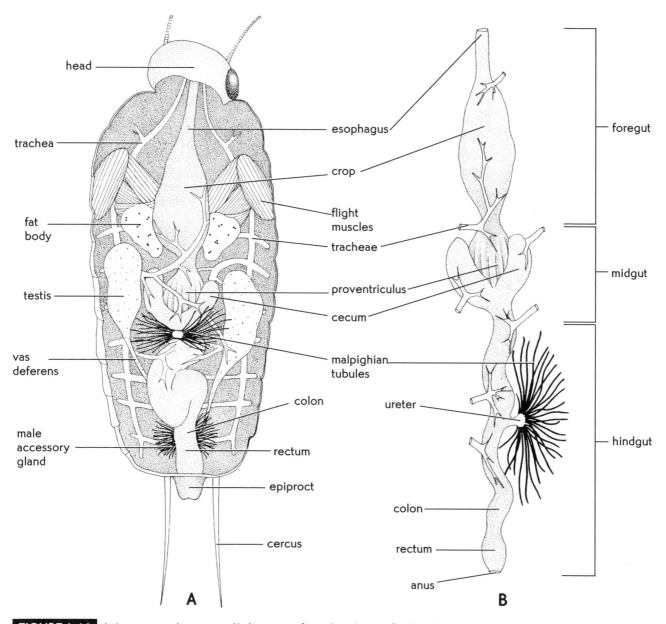

FIGURE 4.13 *Acheta* internal anatomy. A) dissection from dorsal view; B) digestive tract isolated.

Reproductive system. In gravid females the **eggs** are ovulated from the **ovary** into the **oviduct** and retained until the female is fertilized. She can store over 600 eggs in the oviducts, which obscures and compresses all other organs. Therefore, if you have a gravid (unmated) female, remove the ovaries/oviducts before you try to examine any other anatomical features. If the female has mated, the number of eggs is greatly reduced because she lays the eggs as she ovulates them. Identify the reproductive organs in the female or male and look at someone else's dissection to see the other sex. The male produces a complex **spermatophore** which he places in the bursa of the female during copulation. The spermatophore pumps sperm into the **spermatheca** of the female and is then discarded. A male often has a fully formed spermatophore ready, which can be obtained by gentle squeezing of the male's abdomen. If you place a spermatophore on a slide with a drop of water and cover it with a coverslip, it will pump out living sperm which can be observed under a compound microscope.

Digestive and excretory systems. Locate the digestive tract (gut) and observe its placement within the abdominal region. The gut is divided into three main regions: the **foregut,** the **midgut,** and the **hindgut** (Figure 4.13). The foregut consists of the **esophagus,** which connects the mouth to the gut, and the **crop,** a food storage organ. The midgut comprises a hard, round, grinding organ, the **proventriculus,** which is surrounded by two blind pouches which make up the **cecum.** The hindgut is a long tube, the **ventriculus,** which terminates in the **colon** and **rectum.** Near the middle of the hindgut locate the **ureter** and the **malpighian tubules.** The latter two organs are excretory structures which filter and concentrate toxins and impurities from the hemolymph. The malpighian tubules appear yellowish and when removed, can be suspended in cricket Ringers to reveal their tubular structure. With the digestive tract in place note the peristaltic movements, the coiling of the hindgut, and the connections of the tracheae. Grasp the esophagus with forceps and cut just anterior to that point. Slowly lift the anterior end of the gut, carefully cutting the tracheal connections until only the rectum is connected. Cut the rectum and remove the entire digestive tract and lie it beside the cricket in the pan. Place a pin in the esophagus, straighten out the digestive tract, and locate the previously mentioned features.

Nervous system. Leave your specimen pinned in the plate and pour off the saline. The digestive and reproductive systems have been previously removed. If fat body is present, it can be removed by rinsing it out with a stream of Ringer's from a squeeze bottle. Pour off all fluids, add a few drops of methylene blue (or other stain), wait a minute or two, flood with tap water, and observe the stained nervous system. Note the ganglia and nerves.

Cricket Ringer's. To make a stock solution of cricket Ringer's (a saline solution which chemically approximates cricket hemolymph), mix the following ingredients in deionized water:

NaCl	155 Mm	9.10 g/liter
KCl	7 Mm	0.52 g/liter
$CaCl_2.2H_2O$	8 Mm	1.20 g/liter
$MgCl_2.6H_2O$	4 Mm	0.82 g/liter

Apis

The honeybee, *Apis mellifera,* is a highly specialized insect which shows the organized social behavior typical of hymenopterans (Figure 4.14). Bees live in permanent colonies that are partitioned into three **castes,** or class levels. Each caste (the **queen,** the **drones,** and the **workers**) performs certain colonial functions and exhibits different morphologies. The queen, of which each colony has only one, lays the eggs; the drones are males whose only function is to fertilize the queen, and the workers (sterile females which develop from unfertilized eggs) provide most of the other essential functions for the colony. These tasks include hive construction and maintenance, food gathering, and tending to the queen and her young.

External anatomy. Each of the three castes listed previously shows different morphological features. This exercise will concentrate on the anatomy of a worker, but if available, examine a queen and a drone also and note external differences. The body consists of the typical three tagmata, the **head, thorax,** and **abdomen.** The head bears sensory **antennae, compound eyes,** three simple eyes, **ocelli,** and the mouthparts. Place the insect under a dissecting microscope and observe the head features. Locate the chewing mouthparts, the **mandibles,** and the sucking mouthparts, the **labium,** the **labial palps,** and the **maxillae.** The labium is a tongue-like organ with an expanded distal tip; the labial palps lie on either side of the labium; and the maxillae are broad, flattened structures located laterally.

Locate the thorax and examine the membranous **wings.** Note the transparency, the supportive veins, and the locking mechanism. The hindwings are smaller than the forewings but the structure of the two is similar. Now observe the three pair of legs and note the modifications present. The first pair of legs bears a movable spine, the **velum,** at the distal tip. This spine, in conjunction with a more proximal notch, forms the **antenna cleaner.** On the opposite side, locate the **pollen brush,** which consists of a number of stiff bristles. The second pair of legs bears a long **spur** which is used in wax transferal during comb or hive building.

The third pair of legs is highly modified. On the external side, a concave depression bordered by long hairs forms the **pollen basket.** Just below this, on the distal tip of the tibia is a row of spines collectively called the **pecten.** The pecten communicates with the **auricle,** a similar structure on the proximal end of the tarsus. These two structures, collectively called the **pollen packer,** function in the transfer of pollen between the combs and basket. On the inner surface of the hindleg there are nine rows of stiff bristles which make up the **pollen combs.** These combs help remove wax secreted by glands on the ventral surface of the abdomen.

On the abdomen, find a series of pores termed the **spiracles.** These are the external openings to the respiratory system, which in insects consists of a series of ramifying tubules (**tracheal tubes**). At the distal tip of the abdomen, locate the **sting.**

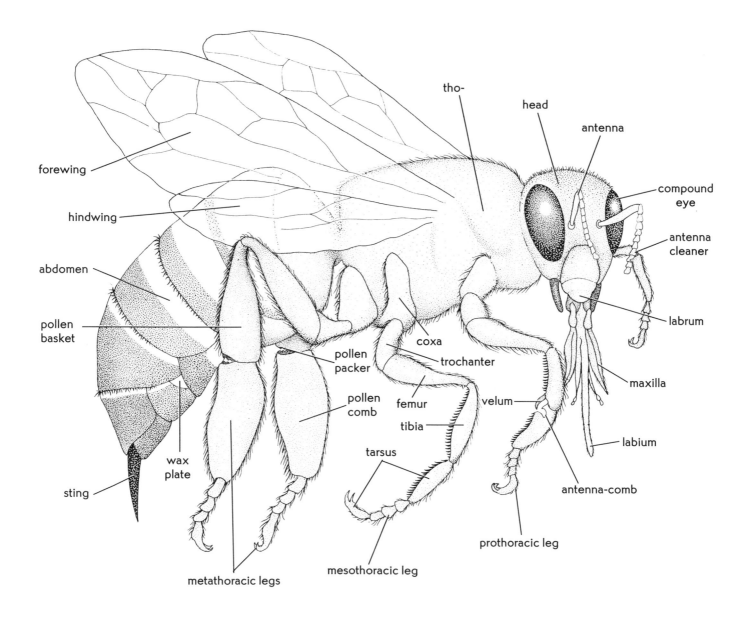

FIGURE 4.14 External anatomy of *Apis,* the honeybee. Specimen shown is a typical worker.

✓ **Checklist of Suggested Demonstrations**

THE ARTHROPODS

_____ 1. Trilobite fossils.

_____ 2. *Limulus* dried specimen.

_____ 3. Preserved spiders (*Latrodectus, Argiope, Tarantula*)

_____ 4. Preserved ticks.

_____ 5. Tick and mite whole-mount slides.

_____ 6. Preserved scorpions.

_____ 7. Pseudoscorpion whole-mount slide.

_____ 8. Whiptail scorpion preserved specimen.

_____ 9. Crustacean naupilus larva slide.

_____ 10. *Cambarus* (or *Procambarus*) preserved specimen.

_____ 11. *Homarus* preserved specimen.

_____ 12. *Emertia* (mole crab) preserved specimen.

_____ 13. *Libinia* (spider crab) preserved specimen.

_____ 14. *Pagurus* (hermit crab) preserved specimen.

_____ 15. *Menippe* (stone crab) preserved specimen.

_____ 16. *Callinectes* (blue crab) preserved specimen.

_____ 17. *Lepas* (goose-neck barnacle) preserved specimen.

_____ 18. *Balanus* (acorn barnacle) preserved specimen or dried shell.

_____ 19. *Eubranchipus* (fairy shrimp) whole-mount slide.

_____ 20. *Bathynomus* (giant isopod) preserved specimen.

_____ 21. Insect order representatives; dried and pinned specimens.

_____ 22. Insect life cycle diagrams and displays.

_____ 23. Insect antennae whole-mount slide.

_____ 24. Insect tracheae whole-mount slide.

_____ 25. Insect compound eye section slide.

_____ 26. Hymenopteran sting whole-mount slide.

Arthropod Notes and Drawings

Arthropod Notes and Drawings

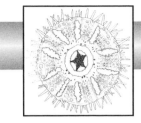

EXERCISE 5

The Echinoderms

THE ECHINODERMS

SYSTEMATIC OVERVIEW

PHYLUM ECHINODERMATA

Marine deuterostomes with bilateral larvae and usually radial adults; a water vascular system, with a unique calcareous endoskeleton; 6,000 living species.

Subphylum Homalozoa

Fossil carpoids, which were not radially symmetrical with rounded or cup-shaped theca and branchials, or arms; attached by stem during all or part of life; oral surface up. Four extinct classes and one living class.

Class Crinoidea

Sea lilies and feather stars; aboral attachment stalk of dermal ossicles; mouth and anus on oral surface; five filter-feeding branching arms with pinnules; ciliated ambulacral groove on oral surface with tentacle-like tube feet for food collecting; spines, madreporite, and pedicellariae absent; 5,000 fossil species and 625 living species (*Antedon, Florometra, Nemaster*).

Subphylum Asterozoa

Radially symmetrical; star-shaped with rays projecting from a central disc; free moving.

Class Asteroidea

Sea stars; arms are not sharply set apart from the central disc; open ambulacra usually have suckered tube feet; aboral surface usually bears pedicellariae; 1,600 living species (*Asterias, Crossaster, Acanthaster, Pisaster, Leptasterias*).

Class Ophiuroidea

Brittle stars and basket stars; arms are sharply set apart from the central disc; ambulacra are closed, covered with ossicles, and contain tube feet without suckers; no pedicellariae; muscular articulated arms, rather than tube feet, provide locomotion; 2,000 living species (*Ophiura, Ophioderma, Gorgonocephalus*).

Class Concentricycloidea

Sea daisies; medusiform asterozoans with a double water-vascular ring and concentric skeletal elements; limited to one species only (*Xyloplax*).

Subphylum Echinozoa

Mostly unattached globoid or discoid without arms; closed ambulacra and suckered tube feet; four extinct and two living classes.

Class Echinoidea

Globular or flattened echinoderms with a rigid test (shell) formed of fused plate-like ossicles; possess movable skeletal spines and several types of pedicellariae; 900 species (*Arbacia, Lytechinus, Strongylocentrotus*).

Class Holothuroidea

Sea cucumbers; cylindrical echinoderms which lack arms, spines, and pedicellariae; microscopic ossicles embedded in leathery muscular walls; ambulacra grooves closed; tube feet with suckers, some modified into tentacles; madreporite plate internal; 1,100 species (*Cucumaria, Eupentacta, Stichopus, Thyone*).

The echinoderms comprise a small (6,000 species), relatively homogenous group of marine invertebrates. Echinoderms are common in all oceans of the world, occurring intertidally and in deep water. The phylum includes the sea lilies, sea stars, brittle stars, sea urchins, sand dollars, and sea cucumbers. Because echinoderms lack an excretory system, they must live in environments which are isotonic relative to their internal body fluids. Thus, echinoderms have never successfully invaded freshwater environments. Phylogenetically, they are placed at the pinnacle of invertebrate evolution because it is thought that they share a common ancestor with the chordates (vertebrates). However, a quick analysis of their anatomy does not reveal advanced structural characteristics, a feature to be expected in animals placed at the top of a phylogenetic scheme.

Echinoderms do share some important developmental features with the chordates, such as method of coelom development (enterocoelous), style of embryonic cleavage (radial), and fate of the blastopore (becomes the anus rather than the mouth). Animals which possess this suite of characteristics are collectively called **deuterostomes**. Although no fossil echinoderms have been found which positively link the phylum with the chordates, they are considered to be related because other advanced invertebrates (molluscs, annelids, and arthropods) show an opposite (**protostome**) style of development. Most zoologists think that the protostome/deuterostome dichotomy represents an early phylogenetic split.

Echinoderms are united by a few common characteristics: they are bilaterally symmetrical as larvae but revert to **radial symmetry** as adults; nearly all species possess a **dermal endoskeleton** made up of calcareous ossicles and spines; and finally, echinoderms possess a hydraulic locomotory system, the **water vascular system,** which is unique to the animal kingdom. Most adult echinoderms show a specialized form of radial symmetry with radiating structures based upon multiples of five. This is termed **pentaradial symmetry** and can be observed in the five rays of most sea stars as well as in the external pore systems of sea urchins and sand dollars. Bilateral symmetry of the larvae is thought to have carried over into the chordates through an evolutionary link.

The endoskeleton is the basis of the phylum name because it imparts a spinous appearance to the skin (echinoderm means "spiny skin"). The water vascular system consists of a series of tubules and passageways which pressurize the system and allow for hydraulic extension and retraction of the primary locomotor organs, the **podia** or **tube feet.** Other features unique to echinoderms include an external respiratory system, the **dermal branchiae** (or **skin gills**), and the presence of small pincer-like **pedicellariae** on the skin surface. The pedicellaria act to clean the skin and deter predation on the dermal branchiae. Primitive features which the echinoderms retain despite their advanced position include the lack of cephalization, an unsegmented body, a poorly-developed nervous system with few sense organs, and a complete lack of excretory organs.

Class Crinoidea: sea lilies and feather stars

Living crinoids (sea lilies and feather stars) represent a small remnant of a once diverse past. An ancient group, crinoids once dominated the seafloors but today are restricted mainly to deep water tropical regions. Crinoids feed on plankton and suspended organic material by ensnaring their prey with mucus-covered tube feet. This feeding style has led some zoologists to postulate that the echinoderm water vascular system evolved as a filter-feeding device. Sea lilies are sessile animals which possess radiating arms attached to a central **stalk.** The arms form a **crown** which is rooted into the stalk by a set of plate-like ossicles, the **calyx.** Feather stars are ecologically similar to sea lilies, but lose the stalk as adults and can swim, creep, or crawl about the seafloor in search of new feeding areas.

Observe some fossil and living species of crinoids on demonstration. Note the branching or feather-like structure of the arms used in feeding. If a fossil specimen is available, locate the stalk and calyx. Many crinoid fossils are represented only by the stalk, because the arms are delicate and easily destroyed or separated during fossilization.

Class Asteroidea: *Asterias*

Class Asteroidea is composed of 1,600 species of sea stars which all possess arms radiating from a central disc. Most sea stars possess five rays, but some species bear 6, 7, 11, or more (up to 50) arms per individual (Figure 5.1). Sea stars are common intertidal organisms often found clinging to hard substrates; some species occur in deep water living in sand or mud. More active at night than by day, sea stars feed primarily on bivalves. Polychaetes, snails, crustaceans, and corals are common prey items also. Many sea stars display bright coloration and conspicuous color patterns which occur in various shades of orange, red, green, grey, and blue. *Asterias* is a typical five-rayed, orange sea star which is commonly found along the Atlantic coast of North America.

LIVING SPECIMENS

Locomotion. Obtain a living specimen of *Asterias* and place it in a glass dish filled with seawater. After the animal has acclimated to its new surroundings, note the slow, methodical movements characteristic of the water vascular system. Sea stars crawl across the substrate on their **oral** (bottom) surface by extending hundreds of

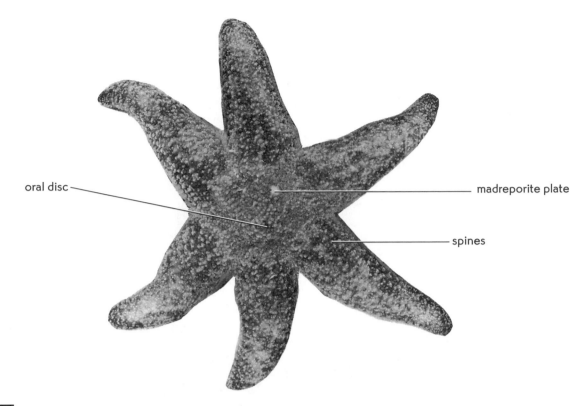

oral disc ———————————————————— madreporite plate

——————————————————————— spines

FIGURE 5.1 An atypical six-rayed sea star, *Leptasterias*. This is a common genus along the Pacific coast of North America.

small **podia (tube feet)** which act as suction cups pulling the organism along (Figure 5.2). The podia lie in the **ambulacral grooves** located on the oral surface of each ray. The tube feet are coordinated for efficiency of movement, but each can be extended or retracted into the ambulacral groove independently. The tube feet are

muscular and filled with fluid; when they contract, water is injected into internal, bulbous storage vesicles, the **ampullae,** which store fluid for the tube feet. When the ampullae contract, water is forced into the tube feet causing them to extend. Observe the extension of the tube feet through the glass dish. Flip the animal onto its

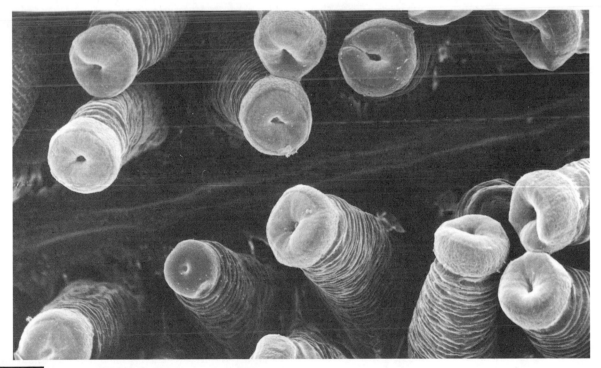

FIGURE 5.2 Tube feet of the six-rayed sea star, *Leptasterias*.　　　　Scanning electron micrograph by T. Das.

aboral (top) surface and note the reaction. The aboral surface is negatively thigmotactic, hence the animal will attempt to turn over or "right itself."

External anatomy. Obtain a preserved sea star and place it in a dissecting pan. Note the **pentaradial symmetry** demonstrated by the presence of the five **arms**, or rays (Figure 5.3). The **central disc** is offset from the rays but not nearly as distinct as in a related type of echinoderm, the brittle star. Note that the animal is flattened along the oral-aboral axis and the **mouth** is located on the oral surface. Cut off a small section of the aboral surface and examine it with a dissecting microscope. The **epidermis** is soft and fleshy and

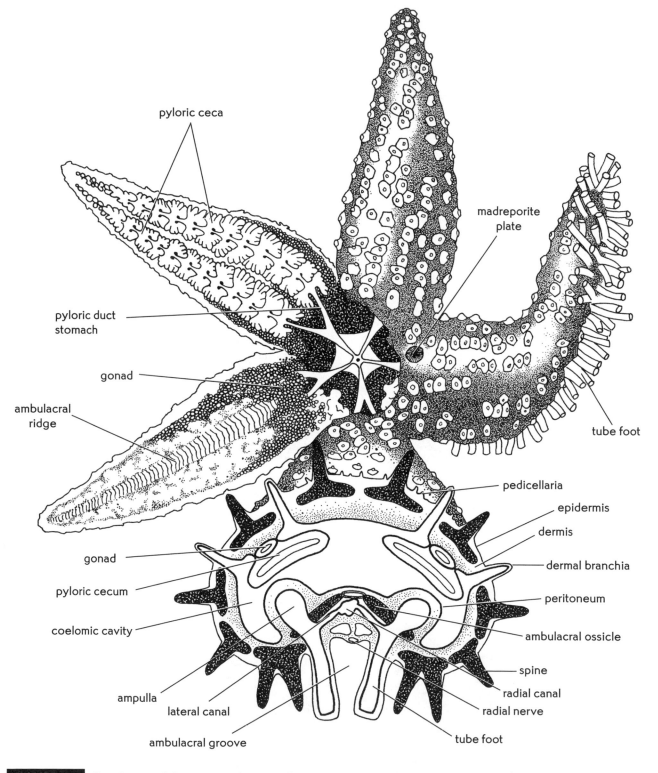

FIGURE 5.3 Aboral view of the sea star, *Asterias*. The rays have been dissected to reveal major internal organs in whole and cross-section views.

protected by calcareous **spines** which protrude through the skin (Figure 5.4). These are extensions of the skeletal ossicles embedded within the skin. Soft, finger-like projections in the epidermis are the **dermal branchiae (skin gills).** These are respiratory organs which are actually membranous extensions of the body cavity. Gas exchange occurs through diffusion into the coelom as long as the epidermis is wet and exposed to dissolved oxygen. Around the base of the spines you will see small **pedicellariae** (Figure 5.5). These are calcareous pincers (modified spines) which capture small prey and protect the dermal branchiae from bacterial colonization, collecting sediment, or parasitic activity.

Aboral surface. The aboral surface of the central disc bears a small, porous sieve plate termed the **madreporite plate** (Figure 5.6). It is a hardened, calcareous structure which regulates the amount of seawater entering the water-vascular system. Note that the two rays which flank the madreporite make up the **bivium;** the other three are associated with the **trivium.** The **anus** is a small central opening on the disc, but it is small and may be hard to locate. Located around the base of each spine is a raised ring of skin bearing the pedicellariae. Some species of sea stars can be differentiated based on the arrangement of the pedicellaria around the spines. Sporadically placed pedicellaria are also located on the surface between the spines. Look at the extreme tip of each ray and determine the location of a small, pigmented **eyespot.** Look carefully! In preserved specimens this sensory structure is hard to identify.

Oral surface. The ambulacral groove of each ray contains four rows of tube feet. The **ambulacral spines** bordering the groove are elongate, movable, and can lock together over the groove to protect the tube feet. Note the size, shape, number, and arrangement of the tube feet within the groove. The podia are used in locomotion but are also important organs in subduing prey items such as bivalves. The hundreds of tube feet, acting in conjunction, generate powerful suction forces which eventually tire and fatigue the valve adductor muscles, thus exposing the soft-parts of the clam. The central **mouth** is surrounded by five pairs of long, movable, protective spines. Note that there is not much room internally to take in a large meal for digestion. To compensate for this, sea stars evert their stomach from their mouth and begin digesting their food externally. This must be done quickly to reduce the exposure time of the viscera. If some tissue is extending from the mouth, it is probably part of the everted **stomach.**

Internal anatomy and dissection. Place the specimen aboral side up in a dissecting pan. Locate the two rays of the bivium and snip off the distal tips. Insert a scissor point laterally into the coelom at the cut end of each ray. Cut toward the central disc along the side of the ray until you reach the disc. Repeat this procedure until you have cut both sides of the two rays. This will allow you to peel up the aboral surface of both rays and reveal the internal anatomy. Cut off the aboral (top) section of the rays and discard it. Now cut around the central disc, staying away from the madreporite and keeping as close to the surface as possible. This will prevent injury to the digestive tract that occupies the central disc. The tissue around the anal opening may cling to the body wall. Pull gently to remove it. A short **intestine** will extend from the anus to the **stomach** (which nearly fills the central disc).

Digestive system. The stomach is divided into two relatively indistinct chambers, the aboral **pyloric stomach** and the oral **cardiac stomach.** The **pyloric stomach** gives rise to a pair of **pyloric ducts** which extend into the arms and connect with a pair of large, lobed **pyloric ceca (digestive glands).** A very short **intestine** leads up from the center of the stomach to the anus in the center of the disc. Below the pyloric stomach is the larger five-lobed **cardiac stomach.** Each lobe of the stomach is attached to the ambulacral ridge of one of the arms by a pair of **gastric ligaments,** which regulates the eversion of the stomach.

As a sea star feeds on a bivalve, it wraps its arms around the animal, attaches its tube feet to the valves, and exerts pressure until the bivalve fatigues and exposes its soft-parts. The animal then contracts its body wall, increases pressure in the coelomic fluid, and everts its stomach. The sea star then inserts the eviscerated stomach into the open shell. There it digests the clam by releasing secretions produced by the pyloric ceca. Partly digested material is drawn up into the stomach

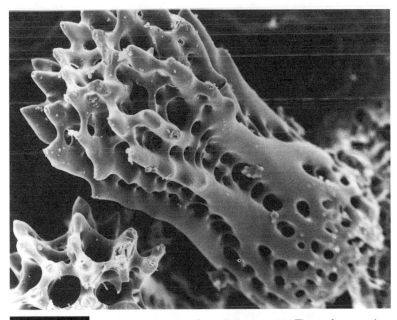

FIGURE 5.4 A calcareous spine from *Leptasterias*. The soft tissue has been removed leaving only the skeletal material intact.

Scanning electron micrograph by T. Das.

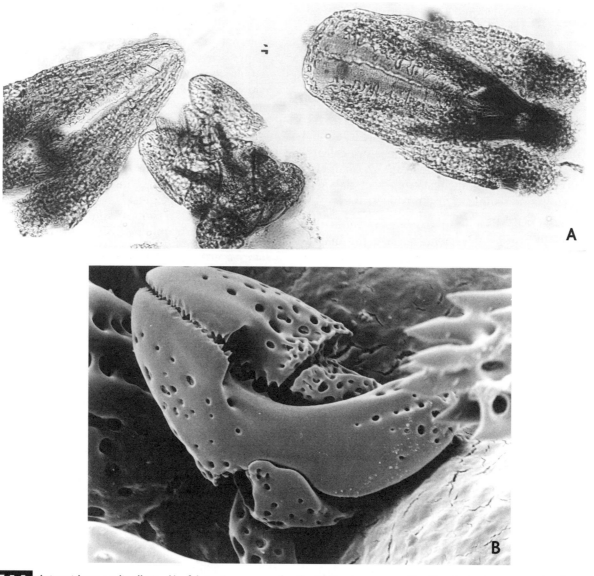

FIGURE 5.5 Asteroidean pedicellaria. A) of *Asterias,* as seen by the light microscope; B) of *Leptasterias,* as seen by the scanning electron microscope. These pincer-like organs help keep the skin clean and free from predation.

Scanning electron micrograph by T. Das.

and pyloric ceca, where digestion is completed. Most echinoderms produce ammonia as an excretory product. There is usually little fecal waste matter. When the sea star is finished with feeding, the stomach withdraws into the coelom by contraction of stomach muscles and relaxation of the body wall; this allows coelomic fluid to return to the arms. Some sea stars feed on small bivalves, such as *Mytilus,* by engulfing the entire animal, digesting out its content, then casting the shell out through the mouth.

Reproductive system. Sea stars are sexually dimorphic, but in preserved specimens it may be very difficult to determine the gender. In one of the dissected rays, remove the lobes of the pyloric ceca and locate the paired **gonads** on either side of the central ambulacral ridge. The gonads emerge from the central disc but

their size varies greatly relative to the reproductive cycle. During the breeding season the gonads are much larger than during nonbreeding periods. If your specimen was captured and preserved during the breeding season the gonads may rival the pyloric ceca in size. Otherwise, the gonads are small and you must carefully examine the base of the ray to locate them. The texture of the gonad is much different from that of the pyloric ceca. The gonads appear orange with a rather bumpy surface. The female gonads may be a little coarser in texture and more orange than the male gonads. Each gonad opens aborally to the exterior at the point of attachment by a very small **genital pore.** Remove a small piece of gonad and smear it across a microscope slide. In the female, the ovary will bear eggs with large nuclei; in the male, the testes will bear many small sperm. When reproducing, large streams of eggs and

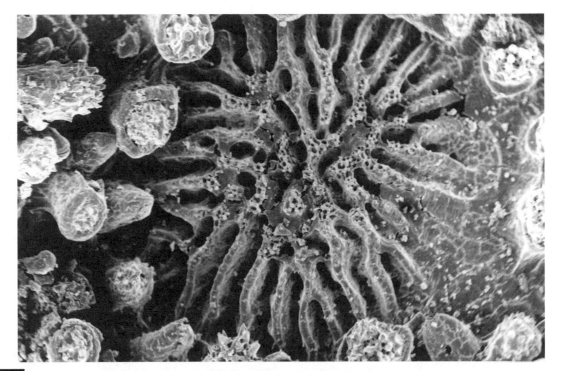

FIGURE 5.6 The madreporite plate of *Leptasterias*. This structure regulates water entering the water-vascular system.

sperm are shed into the water, where fertilization occurs externally. Some species produce pelagic (free-swimming) larvae while others brood the eggs during development.

Dermal endoskeleton. Echinoderms are unique in possessing a dermal endoskeleton formed by binding calcareous plates, or **ossicles,** together with connective tissue. This imparts a protective rigidity to the skin while still allowing some flexibility for movement. Remove the pyloric ceca and gonads (if well-developed) to expose the irregular lining of the endoskeleton. Note the rib-like skeletal network which indicates the presence of the ossicles. Locate the **ambulacral ridge** in the center of each ray. It consists of a conglomeration of tiny ossicles surrounding a series of small pores. The pores allow the passage of fluids from the tube feet to the ampullae. With a needle probe, scrape away a few of the ampullae and locate the pores. Force your needle probe through the pore. If done at the right angle, the probe will pierce an individual tube foot. Press against some of the ampullae and note the elongation effect on the tube feet. Pushing up against the tube feet will cause the ampullae to expand. Both the podia and the ampullae are muscular and can regulate the water pressure by contraction.

Nervous system. The nervous system of echinoderms is primitive and unganglionated. Echinoderms are uncephalized and have little sensory capacity. The sensory system is limited to touch-sensitive cells distributed across the body surface and a handful of light-sensitive **ocelli** located at the tip of each ray. Sea stars possess three interrelated nervous systems. The ectoneural system comprises a network of nerve cell bodies and their processes located just beneath the epidermis throughout the body. The ectoneural system consists of a nerve ring around the mouth in the peristomial membrane; a radial nerve to each arm running along the ambulacral groove to the eyespot; and the nerve plexus, underneath the epidermis. Remove the tube feet and movable spines around the mouth and expose the peristomial membrane. The nerve ring is a whitish thickening on the outer edge of the membrane. To see one of the **radial nerves,** bend an arm aborally and look along the oral surface of the ambulacral groove for a thin white cord. There are two other similar systems; one of which, the **endoneural system,** is located in the oral walls of the cardiac and pyloric stomachs as well as in the lobes of the pyloric ceca. The third system, the **apical system,** is found on the inner surface of the aboral body wall. Both of these are difficult to see and you may be unable to locate them in your dissection.

Water-vascular system. With a pair of forceps remove the stomach from the central disc. Be careful to only remove the soft tissue. Look closely for a small white tube, the **stone canal,** which extends from the **madreporite plate.** The stone canal connects to the inner **ring canal,** which traces the boundary of the central disc above the mouth. Five **radial canals,** one in each ray, radiate out from the ring canal, running along the apex of the ambulacral groove between the ampullae. The canal itself is internal to the groove, but the position of the canal is indicated by the apex of the groove. Short lateral canals (or connecting canals) connect the radial canal with each of the tube feet. Now look on the inside of an arm and study the alternating arrangement of the ampullae. During locomotion, water enters the madreporite plate, circulates down the stone canal to the ring canal, passes out to the rays by way of the radial canals, and finally moves through the lateral canals to the ampullae and the tube feet.

Class Ophiuroidea: brittle stars and basket stars

Brittle stars are fragile, serpentine relatives of the sea stars. The central disc is more pronounced and set apart from the rays than with asteroid sea stars. Nearly 2,000 species of ophiuroids occur in all oceans and at nearly all depths. Many are intertidal or shallow-water animals, where they are found hiding among rocks or buried in mud and seaweed. The arms are more flexible than those of sea stars and lack an open ambulacral groove. The arms do bear tube feet (also called tentacles), but the tube feet lack suckers and are not a primary locomotor organ. The tube feet are chemosensory organs and also function in prey capture. Examine a brittle star, such as *Ophioderma,* and compare the external structure with *Asterias.* Note the unbranched arms and their snake-like appearance. Many species possess elongate, flexible spines which emerge laterally from each ray.

Basket stars possess arms which are highly branched and function in trapping prey. Obtain a dried or preserved specimen of *Gorgonocephalus* and count the number of times each ray divides. In some species the arms bear up to 12 branches. The animal periodically uncurls its arms and captures plankton and small invertebrates which are then transferred to the mouth.

Class Echinoidea: *Arbacia*

The echinoids comprise 900 species of armless echinoderms. The globose sea urchins, the ovoid heart urchins, and the flattened sand dollars are all typical echinoids. They occur in many of the world's oceans in both surface and deep waters where they live among the benthos. Echinoids secrete a protective shell (termed a test) formed of interlocking skeletal plates. The test bears movable spines in all echinoids and long tube feet in sea urchins. Sea urchins are often termed "regular urchins" and sand dollars and heart urchins are called "irregular urchins" because the latter two show some degree of bilateral symmetry even as adults.

Arbacia

The purple sea urchin, *Arbacia,* is a common animal of the North American Atlantic coast. Two other sea urchins, *Strongylocentrotus* and *Lytechinus,* occur on both the Atlantic and the Pacific coasts and can be substituted for *Arbacia* in this exercise. Sea urchins wedge

themselves into cracks and crevices among rocks and coral reefs, feeding on seaweed, algae, coral polyps, and decaying organic matter.

External anatomy. Obtain a preserved sea urchin and examine the test (Figure 5.7). The test, or endoskeleton, consists of ossicles (calcareous plates) that are symmetrically arranged and interlocked or fused to provide an immovable surface. The test grows both by size increase of individual plates and by the production of new plates. The test bears spines attached by means of a ball-and-socket arrangement. Remove a spine and locate the central tubercle which supports it. The spines are movable and operated by a ring of cog muscles. Pull on an attached spine and feel the locking mechanism of the cog muscles. The location of the tube feet is indicated by rows of small perforations along the test. The tube feet extend from these pores to lengths much greater than that of the spines. The arrangement of pores is also used to separate the five ambulacral regions from the interambulacral regions. Study Figure 5.7 and note the different arrangement of spines and individual plates in both regions. The plates are arranged into ten meridional double columns — five double rows of ambulacral plates alternating with double rows of interambulacral plates.

On the aboral surface, note the area that is free from spines, the periproct. The anus is centrally located, surrounded by four (or five) valve-like anal plates. Around the anal plates are five genital plates, each of which bears a genital pore. Note that one of the genital plates is larger than the others and has many minute pores. This is the madreporite plate, a structure which regulates the hydrostatic pressure of the water-vascular system, as in all echinoderms.

Look at the oral surface of the test and locate the central mouth. The edge of the mouth bears five converging teeth and its collar-like lip. The teeth are part of a grinding mechanism called Aristotle's lantern, which is controlled by a set of internal muscles. Examine a dissected specimen with the lantern on demonstration. Note the pentaradial symmetry of the lantern and its inherent rigidity, useful for chewing and tearing seaweed. Some species even use the lantern to carve depressions in rock surfaces to provide suitable hiding places. Surrounding the mouth and lip is a circular membrane termed the peristome. It is perforated by five pairs of chemosensory tube feet called buccal podia and a number of sporadically arranged small spines. Arranged around the peristome are five pair of fleshy gills. The gills open directly into the coelomic cavity and gas exchange occurs through diffusion.

The periostomial region bears a number of pedicellariae on the ends of long, flexible stalks. Smaller pedicellariae are more widely distributed among the spines. Remove a few pedicellariae with a pair of forceps and compare the structure with that shown in Figure 5.5. Now locate the sphaeridia, a series of small modified

A

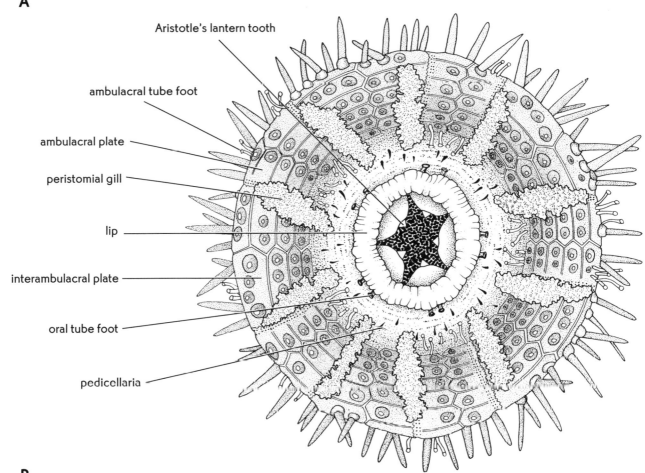

Aristotle's lantern tooth

ambulacral tube foot

ambulacral plate

peristomial gill

lip

interambulacral plate

oral tube foot

pedicellaria

B

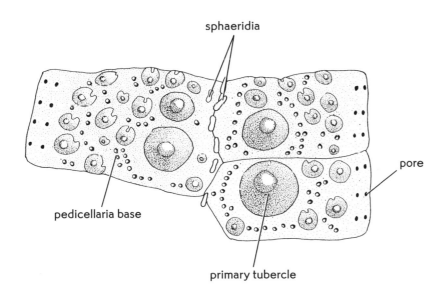

sphaeridia

pore

pedicellaria base

primary tubercle

FIGURE 5.7 External anatomy of the sea urchin, *Arbacia*. A) oral surface; B) detail of several interlocking plates, with spines removed.

spines placed at the junction of the ambulacral plates. Sphaeridia are thought to function in maintaining the proper equilibrium for the animal.

Class Holothuroidea: *Thyone*

The sea cucumbers are somewhat atypical echinoderms because they lack arms, spines, and the rigid skin of most species. Holothuroideans are benthic, slow-moving organisms common in shallow seas, intertidal zones, and most marine environments with muddy or sandy bottoms. The body wall is fairly soft but strengthened with microscopic ossicles embedded in the skin. The tube feet are highly modified. Some are long, flexible and bear suckers; these are used in locomotion. Others form feeding tentacles which surround the mouth. Because many of these animals spend much of their time adhering to the substrate, often only one side of the body bears functional tube feet. Their body symmetry, while still pentaradial, superficially resembles a bilateral arrangement due to an oral-aboral elongation. Some extremely deep water species show a pronounced trend toward bilateralism.

Sea cucumbers, like many echinoderms, are known to possess great powers of regeneration. When disturbed or provoked, sea cucumbers eviscerate their internal organs through the anus, leaving the predator confused and disoriented. Regeneration of the eviscerated organs then occurs. A few species can eject a sticky bundle of tubules which ensnare the predator while the cucumber crawls away. Two common genera found along the Atlantic coast of the U.S. are *Thyone* and *Cucumaria*; *Stichopus* is a familiar genus of the Pacific coast.

External anatomy. Obtain a specimen of *Thyone* (or another suitable genus) and place it in a dissecting pan. The body is elongate with the **mouth** and **anus** at opposite ends (Figure 5.8). The mouth is encircled by branched tentacles which are hollow and connected internally to the water-vascular system. The tentacles are food gathering organs; they are covered with mucus and can extend into the water or over the substrate trapping small prey items, such as plankton. The tentacles are then inserted into the mouth one by one, until the prey is engulfed. The arrangement of the **tube feet** differ drastically between species. Some species, such as *Thyone*, possess podia scattered across the body surface; others, like *Cucumaria*, show the tube feet grouped into five ambulacral regions; still others lack tube feet altogether.

Internal anatomy. Place the sea cucumber on its dorsal surface. The upper concave surface is the dorsal side, and the lower convex surface is the ventral side. With a pair of scissors make a longitudinal cut along the ventral body wall from the anus to the tentacle region. Be sure and stay as close to the body wall as possible. Pin the walls back to the side and note the large **coelomic**

cavity. At the anterior tip, locate the **mouth.** Just below the mouth is the **pharynx** which is supported by a ring of calcareous plates. The pharynx is followed by a short, muscular **stomach** and a long, convoluted **intestine.** The intestine is secured by **mesenteries** and expands at the end to form a **cloaca,** which empties its waste products through the **anus.** Two lateral, branched **respiratory trees** bifurcate from the cloaca; these function in both respiration and excretion. A rhythmic pumping of the cloaca forces air into and out of the respiratory trees, where gas exchange occurs. Several inspirations one minute or more apart are followed by a vigorous expiration that expels all the water. Locate the thin strands that extend from the cloaca to the body wall. These **suspensor muscles** hold the cloaca in place. Extending from the pharynx are a set of long, thin **retractor muscles** that connect to the **longitudinal muscle bands** in the body wall.

The water vascular system consists of a **ring canal** which surrounds the pharynx. Several elongated sacs called **polian vesicles** descend from the ring canal into the coelom. Polian vesicles help regulate hydrostatic pressure within the water-vascular system. A **stone canal** connects the ring canal to the external opening, the **madreporite.** Five **radial canals** extend from the ring canal forward along the walls of the pharynx to give off branches to the tentacles. From there the radial canals run back along the inner surface of the ambulacra, where each gives off **lateral canals** to the **podia** and **ampullae.** Valves in the lateral canals prevent backflow. Note the ampullae along the ambulacra in the inner body wall.

Sea cucumbers are dioecious and fertilization occurs externally. The **gonad** consists of numerous filaments united into one or two tufts which hang from the pharyngeal region. These increase in size with the onset of sexual maturity. A **gonoduct** passes anteriorly to the **genital** pore.

ECHINODERM DEVELOPMENT

Most echinoderms can reproduce both sexually and asexually. Asexual reproduction is often accomplished through regeneration or fragmentation of the body. Sea stars and brittle stars are capable of autotomizing whole arms when attacked by a predator and are known to occasionally split in half. As long as a substantial portion of the central disc is included in the autotomized piece, it can grow into a whole new organism. Sea cucumbers can eviscerate their internal organs to distract predators and quickly regenerate lost parts. Sea urchins probably have the least amount of regenerative powers among a phylum distinguished by this capability, but urchins possess a greater array of defensive spines and armor-like plates than other echinoderms.

Sexual reproduction is common throughout the phylum and most species show separate male and female individuals (although the gender is often quite

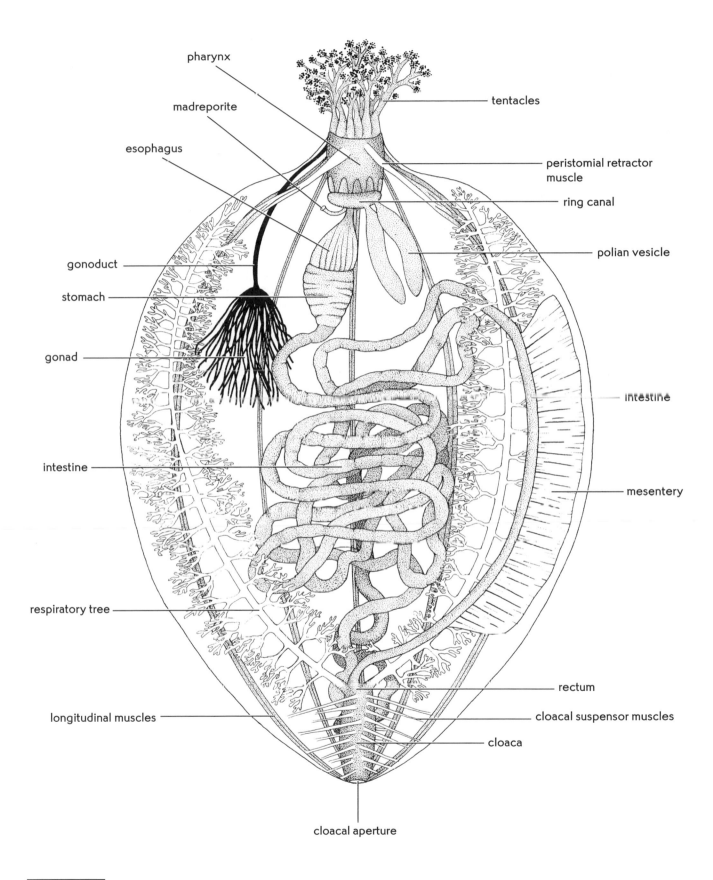

FIGURE 5.8 Dissection of *Thyone,* a sea cucumber.

Labels (clockwise from top):
- pharynx
- madreporite
- esophagus
- gonoduct
- stomach
- gonad
- intestine
- respiratory tree
- longitudinal muscles
- cloacal aperture
- cloaca
- cloacal suspensor muscles
- rectum
- mesentery
- intestine
- polian vesicle
- ring canal
- peristomial retractor muscle
- tentacles

difficult to distinguish). During spawning season, adult echinoderms release huge numbers of eggs and sperm into the water column where fertilization occurs. Nearly half of all echinoderm species brood their eggs on the body surface of the female or in a specially formed brooding chamber. The remaining echinoderm species release their gametes and produce planktonic larvae which are easily dispersed to new areas of habitat.

Echinoderms show the typical **deuterostome** pattern of development: **radial** and **indeterminate cleavage, enterocoelous** coelom formation, and the anus is formed from the **blastopore.** In sexual reproduction, the gametes fuse to form a **zygote,** or fertilized egg (Figure 5.9). The zygote then passes through several stages of **cleavage** as it progresses from an embryonic 2-cell, 4-cell, 8-cell, 16-cell, and finally to a 32-cell stage. These early stages form through mitosis (nuclear division) but there is no associated cell growth, as in normal somatic cell mitosis. Instead, with each subsequent division the cells are reduced in size by one half. This cleavage process transforms a large, single-celled egg into a solid mass of smaller, easily maneuverable cells (each of which is called a **blastomere**). Eventually, this solid mass differentiates into a hollow, spherical ball of cells. This represents the **blastula** stage and is characterized by possessing a fluid-filled central cavity, the **blastocoel.** At this stage in development, the embryo consists of several hundred to a few thousand cells. The following stage, the **gastrula,** is characterized by a series of important events. First, some of the cells begin to push in from the surface of the embryo, creating a small depression, in a process termed **invagination.** An external pore appears, the **blastopore,** from which cells extend internally to create the **archenteron,** the precursor to the digestive tract.

In echinoderms and chordates, the blastopore becomes the anus and the mouth develops late in gastrulation. In molluscs, arthropods, and annelids (protostomes), the blastopore forms the mouth. The gastrula now consists of two embryonic **germ layers,** the **endoderm** and the **ectoderm.** The endoderm will give rise to the lining of the digestive tract; the ectoderm will develop into the nervous system and the external structures of the body, such as the skin. A third germ layer, the **mesoderm,** penetrates the previous two layers and gives rise to the organs and cells which make up the vast majority of the adult body. The developmental stages mentioned in this paragraph are common to nearly all animal species. Developmental patterns and the emergence of larval forms which are specific to different phyla begin to appear after gastrulation.

Echinoderms show several larval forms which are associated with the different subphyla and classes. The first larvae of sea stars and other asteroids are planktonic filter-feeders which develop from the gastrula. These **bipinnaria** are characterized by bands of locomotory cilia and the presence of a complete gut (Figure 5.10). As mentioned previously, the larval forms of echinoderms are bilaterally symmetrical; the adults then revert to a specialized form of radial symmetry. The bipinnaria eventually grow arm-like lateral extensions from the body wall which distinguish the next

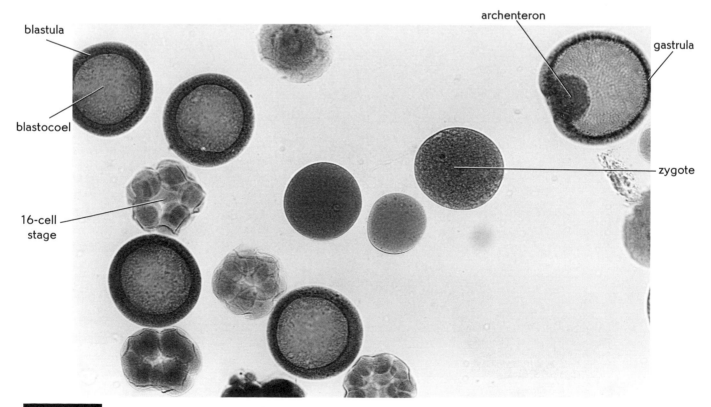

FIGURE 5.9 Composite micrograph of typical animal developmental stages from the unfertilized egg to the gastrula.

larval form, the **brachiolaria** (Figure 5.10). The brachiolaria live pelagically for a short while, then settle to the ocean floor and attach to the substrate. Here they lose their larval arms and begin metamorphosing into the typical adult form. Other echinoderms, such as the brittle stars and the echinoids, show a slightly different larval stage, the **pluteus**, which develops from a generalized **dipleurula** larva (Figure 5.10). Examine a series of prepared slides which show the progression from zygote to larval form and compare these to the micrographs shown in Figures 5.9 and 5.10.

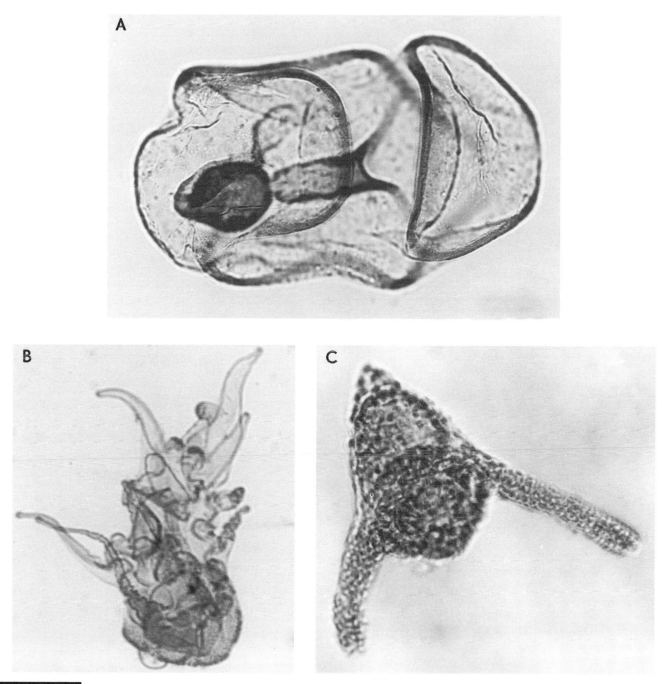

FIGURE 5.10 Echinoderm larval forms. A) bipinnaria; B) brachiolaria; and C) pluteus.

✓ Checklist of Suggested Demonstrations

THE ECHINODERMS

_____ 1. Echinoderm early development slide.

_____ 2. Bipinnaria larva whole-mount slide.

_____ 3. Brachiolaria whole-mount slide.

_____ 4. Preserved and fossil crinoids.

_____ 5. Assorted species of dried asteroidean sea stars.

_____ 6. Asteroidean sun star dried specimen.

_____ 7. *Asterias* pedicellaria whole-mount slide.

_____ 8. *Asterias* ray cross-section slide.

_____ 9. Pluteus larva whole-mount slide.

_____ 10. Assorted dried specimens of ophiuroidean brittle stars.

_____ 11. *Gorgonocephalus* dried or preserved specimen.

_____ 12. Various sea urchin, sand dollar, and heart urchin dried specimens.

_____ 13. Sea urchin dried tests.

_____ 14. *Arbacia* preserved specimen.

_____ 15. Aristotle's lantern.

_____ 16. *Dendraster* preserved specimen.

_____ 17. Heart urchin preserved specimen.

_____ 18. Various sea cucumbers (*Thyone, Stichopus*).

Echinoderm Notes and Drawings

Echinoderm Notes and Drawings

Echinoderm Notes and Drawings

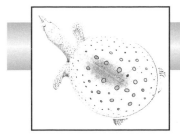

EXERCISE 6

The Hemichordates and Chordates

SYSTEMATIC OVERVIEW

PHYLUM HEMICHORDATA

Acorn worms; worm-like animals characterized by having bilateral symmetry, a well-developed enterocoelom, gill slits, as well as a primitive dorsal nervous system and an internal skeleton in the form of a notochord; 120 species (*Balanoglossus, Dolichoglossus, Saccoglossus*).

PHYLUM CHORDATA

Animals possessing, at some stage in their life cycles, well-developed gills or gill slits, a dorsal hollow tubular nerve cord, and a notochord; 47,200 species.

Subphylum Urochordata

Tunicates; having a larval stage in which chordate characteristics are present, the neural tube and notochord being lost in the sedentary adult, although it possesses a primitive circulatory system; 1,400 species (*Ciona, Molgula*).

Subphylum Cephalochordata

Lancelets; having a well-developed coelom, a circulatory system without a discrete heart, and a fusiform body that has prominent muscle segments (myotomes); 30 species (*Amphioxus* [*Branchiostoma*]).

Subphylum Vertebrata

Vertebrates; having a cranium (skull), visceral arches, and a spinal column of segmented vertebrae that are cartilaginous in lower forms and bony in higher forms; notochord extends from cranium to base of tail; enlarged brain; head region with specialized sense organs; 45,710 species.

Class Agnatha

Cyclostomes; having a long, slender, and cylindrical body with median fins, a mouth located ventrally, 5—16 pairs of gill arches and a persistent notochord, but lacking true jaws and scales; 60 living species (hagfish and lampreys).

Class Chondrichthyes

Having a cartilaginous skeleton with notochord, tough skin that is covered with scales, median and paired lateral fins, a ventrally located mouth with both upper and lower jaws, and pectoral and pelvic girdles; 650 living species (sharks, skates, and rays).

Class Osteichthyes

Having a skeleton that is somewhat bony, a terminal mouth, gills covered by an operculum, median and paired fins, and, usually, skin that is covered with scales, although some are scaleless; 21,000 species (perch, carp, trout, and many others).

Class Amphibia

Having moist, glandular skin that lacks scales; two pairs of limbs, but no fins; a bony skeleton; a terminal mouth with upper and lower jaws and a tongue that is often protrusible; aquatic in the larval stage but usually terrestrial as an adult; 3,900 species (frogs, toads, salamanders, and newts).

Class Reptilia

Having a body that is dry and covered with scales; two pairs of limbs (absent in snakes), with digits adapted to running, crawling, climbing, or swimming; and a bony skeleton; 7,000 living species (lizards, snakes, turtles, crocodiles, and alligators).

Class Aves

Warm-blooded animals having a body covered with feathers, forelimbs modified as wings, a bony but light skeleton, and a beak; 8,600 species (includes all birds).

Class Mammalia

Having mammary glands that secrete milk for nourishing the young, hair in varying quantities, and young that are born alive; 4,500 species (rodents, cats, dogs, horses, seals, whales, humans, and many others).

THE HEMICHORDATES

The hemichordates are a group of small, soft-bodied marine animals. They are common animals of broad distribution often found in U-shaped burrows on sandy or muddy substrates. They may live singly or in colonies, where their sluggish movements help draw food into their burrow. Because they possess both echinoderm and chordate-like characteristics, the hemichordates presumably represent an evolutionary link between the echinoderms and the chordates. In fact, hemichordates were historically classified as chordates until the 1940's. There are two basic types of hemichordate: an elongate worm-like group, the **acorn worms;** and a group of sedentary, vase-shaped animals, the **pterobranchs.** We will concentrate on the acorn worms as representative examples. There are three very similar genera of acorn worms; *Balanoglossus, Dolichoglossus,* and *Saccoglossus.* Any of the three will suffice for this exercise.

Examine a preserved specimen of either *Balanoglossus, Dolichoglossus,* or *Saccoglossus.* Note the flexibility of the body and its elongate form (Figure 6.1). Different species range in size from a length of 10 cm to nearly 1.5 m, but all retain a similar body plan. The body is divided into three distinct regions: the anterior **proboscis,** the short **collar,** and a long **trunk.** The anterior end of the trunk bears a series of respiratory pores which allow water to enter the **pharynx.** There can be as few as 10 pairs or as many as 100 pairs of these **gill slits,** depending upon the species. Distally, the trunk bears a **gential region** and an **abdominal region.** The epidermis is ciliated and covered with mucus-secreting cells, which are useful in burrowing and feeding. The body cavity is reduced and filled with connective tissue and muscle. The muscle tissue of the proboscis and collar is better developed than that of the trunk. Movement is accomplished by contraction of the anterior end, while the trunk is pulled along the substrate passively.

Acorn worms can reproduce both sexually and asexually. Asexual reproduction can be accomplished through regeneration; damaged worms can reform lost parts and new individual worms can form from pieces of the trunk. The adults are dioecious, but the males and females superficially resemble each other. The **gonads** occur in rows along the trunk; fertilization is external as gametes are shed through external pores into the sea water. Often there is no larval form, but some species pass through a larval stage, the **tornaria,** which resembles the bipinnaria larvae of echinoderms. The adults possess a suite of chordate-like features, such as external pharyngeal gill slits and a dorsal hollow nerve cord. They also show some invertebrate characteristics, like a solid ventral nerve cord, and an annelid-like circulatory system.

Most acorn worms are filter-feeders which digest organic remains out of the benthos and draw plankton into their burrows. As they filter-feed, acorn worms wave the proboscis, causing sediment and plankton to stick to the epidermis. Cilia then drive food into the mouth. The collar can move up or down to regulate the amount of food, or the size of particles, that enter the mouth. Acorn worms possess an open circulatory system which fills the coelom with **hemolymph.** Oxygen diffuses into the hemolymph through the gills and the skin. The nervous system is primitive and few sensory organs are present. A nerve net lies under the epidermis and communicates with a dorsal and ventral **nerve cord.** The nerve cord primarily transmits impulses to the collar to allow rapid open and closing of the mouth. Sensory cells are distributed across the proboscis.

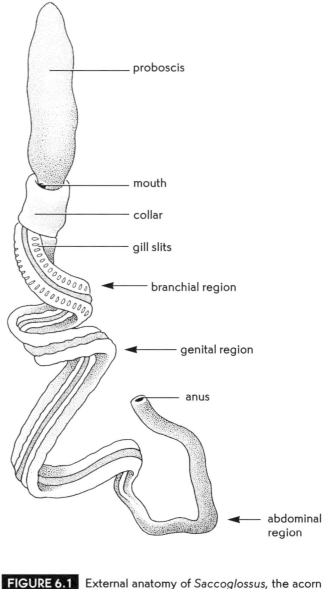

proboscis

mouth

collar

gill slits

branchial region

genital region

anus

abdominal region

FIGURE 6.1 External anatomy of *Saccoglossus,* the acorn worm.

THE CHORDATES

The phylum Chordata contains some of the most diverse and ecologically significant animals on earth. An incredible diversity of form and complexity of structure and behavior have combined to form some of the most successful and dominant organisms ever known. From the ancient armored fishes (placoderms), to the prehistoric dinosaurs and their flying descendents (birds), and to the giant mammals of a more recent past, the chordates have ruled the earth since their inception. The phylum, while limited in number of species (approximately 47,000), contains organisms that have conquered all types of aquatic and terrestrial habitats. The phylum consists of three subphyla; the **Urochordata** (tunicates), the **Cephalochordata** (lancelets), and the **Vertebrata**. Of these, about 2,000 are filter-feeding marine invertebrates (the Urochordates and Cephalochordates). The remaining 45,000 species comprise a well-known and highly conspicuous group of animals, the vertebrates.

Most organ systems of chordates are well-developed, but it's the efficiency and expansion of the nervous system that has been chiefly responsible for the success of the chordates. Phylogenetically, chordates are lumped into a single group based on the presence, at some point in the life cycle, of four major features: the **notochord, pharyngeal gill slits, a dorsal tubular nerve cord,** and a **post-anal tail.**

Notochord. The notochord is a thin, cartilaginous rod lying dorsal to the gut and extending most of the length of the animal. It functions as a support organ, strengthening and imparting rigidity to the organism. In the lower chordates (the uro- and cephalochordates), it persists as a functional structure throughout the life of the organism. In higher chordates (vertebrates), it appears in the embryo where it is surrounded and later persists as the soft center of the intervertebral discs.

Pharyngeal Gill Slits. The pharyngeal gill slits are a series of paired slits in the pharynx, which serve as passageways for water to the gills or as a filter-feeding organ. Lower vertebrates retain them throughout the life cycle; higher vertebrates possess them only as embryonic structures. In many vertebrates, the gill slits never completely form openings, but develop into pouches that appear to have no function.

Dorsal Tubular Nerve Cord. The dorsal tubular nerve cord and its anterior expansion, the brain, combine to form an extremely well-developed nervous system. The nerve cord is hollow, fluid-filled, and lies dorsal to the alimentary tract. The nerve cord is a primary organ of all chordates and is retained throughout the life cycle of all species.

Post-anal Tail. The tail projects beyond the anus at some stage in the life cycle and functions as a locomotory or balancing organ in some species. It may or may not persist in the adult, depending upon need. Some aquatic animals retain the post-anal tail as a propulsion organ; arboreal (tree-living) animals use the tail for balance and as a prehensile grasping organ; and many terrestrial quadrupeds use the tail to increase agility and balance as they run.

SUBPHYLUM UROCHORDATA: *CIONA* AND *MOLGULA*

The urochordates are a group of sessile marine organisms that are covered by a firm, leathery covering, termed the **tunic.** The tunic is actually composed of **tunicin,** a cellulose-like molecule which imparts support and protection to the body. Tunicates are unique among the chordates in lacking a coelom and vestiges of metamerism during larval or adult stages of life. In general, tunicates do not resemble other chordates and have many features analogous to those of other sessile animals. Most species occur in water less than 100 meters deep, but about 100 of the 1,400 known species descend to depths of 5,000 meters. A few species inhabit beaches, living among the grains of wet sand. The subphylum contains 3 classes: the Ascidiacea (sea squirts), the Larvacea (appendicularians), and the Thaliacea (salpians). Of these classes, the ascidians are by far the most successful group (1,300 species) and they typify the subphylum. The larvaceans and the thaliaceans are free-swimming planktonic animals. The sea squirts are generally small, translucent animals which live attached to hard substrates or in sand and mud. The adults are sac-like or cylindrical with two prominent siphons used in filter-feeding.

External and internal anatomy. Examine a preserved specimen of the sea squirt *Ciona* (Figure 6.2). Note the outer body wall, the **mantle,** is comprised of several layers. The outer **epidermis** secretes the tunic and overlies the **dermis** and three muscular layers. Immediately visible are the **excurrent siphon** and the **incurrent siphon,** both of which allow water to circulate through the body and can be closed by muscular action. The incurrent siphon is located at the free tip of the body; the excurrent siphon usually is oriented more laterally. The body is divided into an anterior **atrial cavity,** containing a large pharyngeal sac that bears numerous **gill slits,** and a more posterior **visceral cavity,** which houses the stomach, heart, and reproductive organs. While filter-feeding, cilia located near the gill slits draw water and nutrient particles into the body and through the **mouth.** Water passes through the pharyngeal gill basket into the surrounding atrial cavity and exits through the excurrent siphon. Food particles are trapped in a mucosal layer of the pharynx, which is secreted by the endostyle. Ciliary action then drives the nutrients toward the stomach where digestion occurs. The **intestine** ascends into the atrial cavity where waste products are expelled by the **anus** near the excurrent siphon. A small tubular **heart** pumps hemolymph into and out of the body sinuses in an alternating pattern. The hemolymph transports dissolved gases and body wastes in addition to bathing the organs in a nourishing fluid.

The nervous system is limited to a nerve net and single ganglion in the mantle and sensory structures are nearly absent. Some sensory cells are located near the opening of the siphons. Each individual sea squirt possesses functional **testes** and **ovaries** located in the visceral cavity. A pair of **genital ducts,** which service each gonad, pass into the atrial cavity and open near the excurrent siphon. Gametes are shed externally and fertilization occurs in the open water.

Tadpole larva. The larva of many tunicates superficially resembles the larval form of vertebrate frogs, and hence is called the **tadpole larva** (Figure 6.3). Examine a prepared slide of the larval stage of *Molgula* or another suitable tunicate. *Molgula* is a sea squirt similar to *Ciona* in structure and behavior, but lacks the elongate appearance of *Ciona*. The tadpole larva is transparent and planktonic. The larva possesses the four general chordate characteristics, but only the gill slits are retained into adulthood. Its tail contains a supporting notochord, gill slits in the pharnyx, a dorsal hollow tubular nerve cord, and serial pairs of lateral, segmental muscles. During development, it swims freely for a time, then settles down on the substrate, and transforms into a sedentary adult.

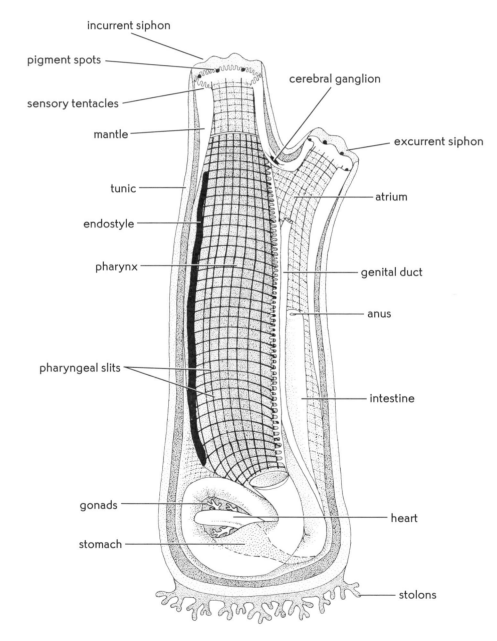

FIGURE 6.2 Internal anatomy of *Ciona*, the solitary sea squirt.

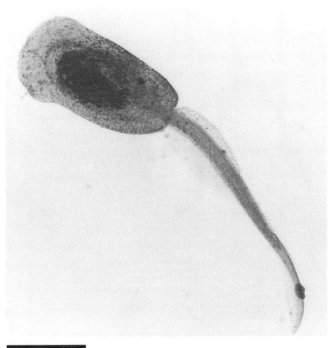

FIGURE 6.3 The tadpole larva of *Molgula*, a typical tunicate.

SUBPHYLUM CEPHALOCHORDATA: AMPHIOXUS (BRANCHIOSTOMA)

Most cephalochordates are streamlined animals which are commonly called **lancelets** due to their piercing shape. *Amphioxus* (*Branchiostoma*) is a small, fish-like animal found in shallow marine waters throughout the world. These animals can swim with lateral, undulating movements of its body, but they spend most of the time buried in the sandy substrate with their anterior end projected. They are most common in smooth, porous sands which harbor little organic matter. *Amphioxus* is a typical cephalochordate and possesses the four chordate characteristics throughout its life cycle.

External and internal anatomy. Examine prepared slides of *Amphioxus* in both whole-mount and cross-section views (Figure 6.4). The animal is somewhat translucent and many internal features are visible in a stained whole-mount slide. Note the conspicuous chevron-shaped **muscle bands** (**myotomes**) which extend across much of the body. A median **dorsal fin** runs along the top of the body and a median **ventral fin** covers the posterior third of the animal. The dorsal fin is supported by a series of short rods of connective tissue called **fin rays.** Running directly underneath the fin rays is the **nerve cord,** which extends nearly the entire length of the body. Ventral to the nerve cord is the supportive **notochord,** which helps prevent injury or bending of the nerve cord.

The digestive tract lies ventral to the notochord. The **mouth,** located in the anterior **buccal cavity,** leads to the elongate **pharynx.** The pharynx consists of a series of **gill slits** located between the **gill bars** in the wall of the pharynx. The pharynx is ciliated and produces currents that beat inward to supply the animal with water. This water enters the mouth, passes over the gill slits (where suspended food particles are trapped), and exits the body through the **atriopore,** an external opening at the end of the **intestine.** Toward the posterior end of the pharynx, a lobe-like outgrowth, the **liver,** emerges. The liver releases digestive enzymes into the intestine. The **gonads** are located between the liver and the atriopore, just ventral to the intestine.

The anterior end of the body houses the funnel-shaped buccal cavity which is surrounded by a circle of external tentacles called **cirri** (Figure 6.5). Within the buccal cavity a set of smaller internal organs, the **oral tentacles,** encircle the mouth. Note the snout bears the anterior tips of the notochord and the nerve cord.

Obtain a cross-section slide of *Amphioxus* taken from the pharyngeal region of the animal. Locate the following structures: the one-celled outer covering, the **epidermis,** and the thicker **dermis;** the **dorsal fin** and the supportive **fin ray;** a series of lateral muscle bundles, the **myotomes;** the **nerve cord** and the underlying **notochord;** some small pockets which represent the **coelom;** a small **dorsal aorta;** a large, open **pharynx** which contains the gill bars and slits; a section of the digestive **liver;** and **gonads** on either side of the **atrial cavity.** The **gonads** are paired bodies containing gametes. *Amphioxus* is dioecious and the sex of the specimen can be determined by comparing the cells of the gonads. Males possess testes containing a large number of small, dense cells. Females bear ovaries characterized by fewer, larger cells with vesicular nuclei. Near the ventral surface identify the **ventral aorta** which is surrounded by the **endostyle;** and a pair of external **metapleural folds.**

SUBPHYLUM VERTEBRATA

Vertebrates are by far the most successful, advanced group of chordates. In fact, vertebrates show such an intriguing array of structural complexity and behavioral adaptations that many of our scientific inquiries often focus upon the vertebrates. The name of the subphylum refers to the presence of metameric **vertebrae** (backbones) which support and protect the fragile nerve cord. In sum, the vertebrae comprise the **spinal column,** a linear support system which allows flexibility of the trunk while imparting strength and stability to the body. The spinal column articulates with the **skull,** a hollow bony case which protects the brain.

Vertebrate evolutionary history extends back over 600 million years. During this time several different groups have risen to prominence, only to be replaced as the dominant species by the next successful group of vertebrates. There are currently seven classes of living

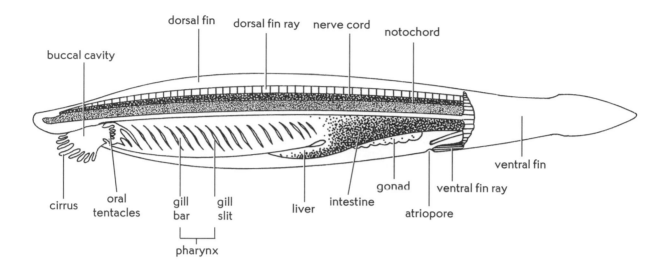

A

buccal cavity

dorsal fin

dorsal fin ray

nerve cord

notochord

cirrus

oral tentacles

gill bar

gill slit

liver

intestine

gonad

atriopore

ventral fin ray

ventral fin

pharynx

B

dermis

dorsal fin ray

central canal

dorsal aorta

hyperbranchial groove

liver

atrial cavity

ventral aorta

epidermis

dorsal fin

nerve cord

notochord

coelom

myotome

pharynx

gonad

endostyle

metapleural fold

FIGURE 6.4 *Amphioxus (Branchiostoma)*, a lancelet. A) lateral view of internal anatomy; B) cross-section through pharyngeal region.

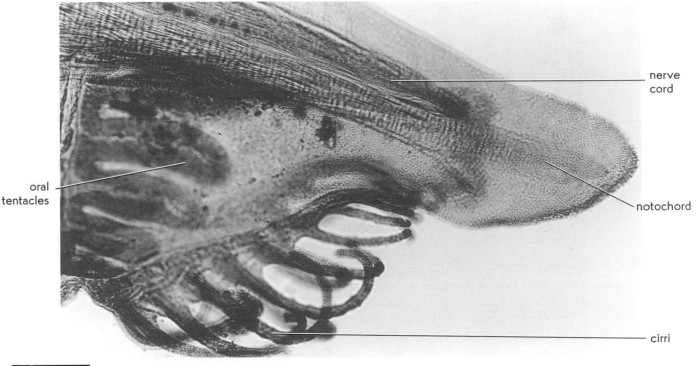

oral tentacles

nerve cord

notochord

cirri

FIGURE 6.5 The anterior end of *Amphioxus*. Note the characteristic features from Figure 6.4.

vertebrates: **Agnatha, Chondrichthyes, Osteichthyes, Amphibia, Reptilia, Aves,** and the **Mammalia.** Two primitive classes, the **Placodermi** and the **Acanthodii,** are extinct groups known only from fossil specimens. Of these two groups, the placoderms (armored fishes) were more successful, dominating the seas around 350 to 400 million years ago. Most placoderms species averaged 20–30 cm in length, but a few gigantic species reached lengths of 10 m. These were incredibly heavy fish, for placoderms possessed dense dermal plates which protected the head and anterior portions of the body. Placoderms were probably bottom dwellers, scavenging food or attacking small fish.

Class Agnatha: *Petromyzon*

Agnathans are jawless fishes which are often referred to as **cyclostomes** ("round mouths") because they have no jaws to open and shut the mouth. As a consequence, the agnathans possess a permanently open, circular mouth. Agnathans evolved before jaws were developed out of an anterior pair of gill arches, hence they are the most primitive living vertebrates. This group dominated marine environments from 400 to 500 million years ago; today they are limited to about 60 species, most of which are marine. Brook lampreys, sea lampreys, and hagfishes are common living examples. The hagfishes are entirely marine, whereas lampreys inhabit both marine and freshwater. In addition to the lack of true jaws, agnathans bear some primitive features such

as, a cartilaginous endoskeleton, a complete lack of external scales, a lack of paired appendages, and adult retention of the notochord.

Examine the sea lamprey, *Petromyzon*, on demonstration (Figure 6.6). Note the elongate shape of its body, and lack of paired fins. The lamprey does possess **median fins** including **anterior** and **posterior dorsal fins** and a circumventing **caudal fin.** The lamprey is cephalized and the anterior end bears seven **external gill slits,** an image-producing **eye,** and a single **median nostril.** On the ventral surface of the head, locate the **mouth** which lies within a suctorial disc called the **buccal funnel.** Within the buccal funnel, lampreys possess a number of **horny teeth** which are arranged in a circular pattern surrounding the rasping **tongue.**

Although some small species of lamprey are filter-feeders, most large species are predatory animals which attack fish. They swim rapidly through the water toward the prey item and use the buccal funnel and sharp teeth to attach to the flesh of their prey. Then, using the teeth and the rasping tongue, the lamprey wounds the fish, secretes an anticoagulant, and feeds on blood and tissue fluids. Lampreys generally remain attached to the host until they are engorged with blood, then dissociate from the fish, leaving an open, festering wound that is easily infected with bacteria and fungi. Large populations of lampreys can greatly deplete fish populations, especially large, commercial fish which are important food sources for human culture. Chemical and electrical barriers are known to be successful in blocking the spawning migration of the adults (into freshwater) and hence have controlled the populations in many ecologically sensitive lakes and streams.

Class Chondrichthyes: *Squalus*

The class Chondrichthyes includes a variety of fish with flexible **cartilaginous skeletons,** a ventral **mouth,** and 5–7 pairs of ventral **gill slits.** Most chondrichthyan fish are marine organisms, but some species live in brackish or freshwater. Many extinct species inhabited freshwater. The class includes sharks, skates, rays, ratfish and sawfish (Figure 6.7). All species are predatory and many are active swimmers feeding upon a variety of marine invertebrates and vertebrates, such as squid, fish, crustaceans, and even some large marine mammals.

While the skeleton is mainly cartilaginous, some parts of the skull and the vertebrae are reinforced with hard calcium deposits. This imparts some rigidity to the axis of the skeleton without compromising the inherent flexibility of cartilage. The skin is rough and thick, providing protection from surface wounds. The skin is covered by small, pointed **placoid** scales made of plates of dentine covered by enamel; this is similar in structure to vertebrate teeth. In fact, the "teeth" of sharks are actually modified placoid scales which are routinely

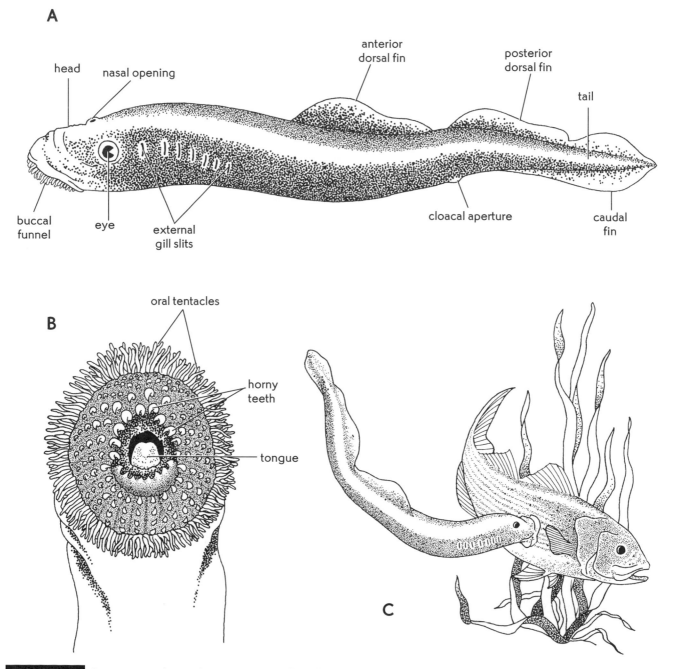

FIGURE 6.6 *Petromyzon*, the sea lamprey, an agnathan. A) lateral view showing external features; B) buccal funnel; C) lamprey preying on fish.

shed and replaced. A typical adult shark may form and shed as many as 40,000 teeth in a lifetime.

Chondrichthyan fish lack a functional swim bladder to keep them buoyant and many species must keep in constant swimming motion to stay afloat and provide clean water to aerate the gills. Most fish in this class possess a **heterocercal tail**, in which the dorsal lobe of the caudal fin is larger than the ventral lobe. This arrangement aids the fish by producing an uplifting swimming force, keeping the animal elevated in the water column. In addition, chondrichthyans also possess a **spiral valve**, an internal coil of absorptive tissue in the intestine.

Sharks are perhaps best known for their occasional tendency to attack human swimmers and divers. Obviously, humans are not a typical prey item for most sharks, but some species of shark which regularly prey on large fish and marine mammals do not hesitate to attack ungainly human swimmers, especially when crowds of people are flopping about in shallow waters. The great white shark (*Carcharodon carcharias*) is perhaps the most infamous due to its large size (6–7 m) and apparent viciousness, but the largest of all sharks,

the whale shark (*Rhincodon typus*), which reaches a length of 18 m, is a filter-feeder which strains great quantities of plankton.

Skates and rays are quite different in form and behavior from a typical shark. Their pectoral fins are greatly expanded and their bodies are flattened dorsoventrally to provide an efficient gliding movement. They often hover just inches above the ocean floor in search of small invertebrate prey.

Sawfishes are known for their elongate skull which terminates in a many-toothed "saw." The snout is used to slash through dense schools of fish, maiming and injuring individuals that are subsequently ingested.

An unusual group of chondrichthyan fish, the chimaeras or ratfish, are probably only distantly related to the other forms. They possess a long, rodent-like tail, and several features which diverge from the typical chondrichthyan plan. They are mainly bottom dwellers that feed on molluscs and small crustaceans.

Squalus

Squalus acanthias, the dogfish shark, is a relatively small species common in coastal waters of the Atlantic

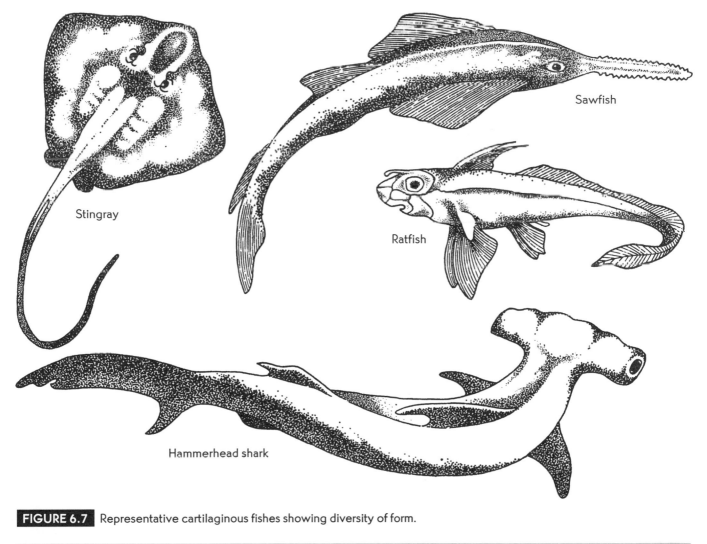

FIGURE 6.7 Representative cartilaginous fishes showing diversity of form.

and Pacific. This species travels in schools and is known to make migrations which cover thousands of miles. Obtain a perserved specimen of *Squalus* and locate the external features labeled in Figure 6.8. The body is divided into the anterior **head,** the median **trunk,** and the posterior **tail.** The bluntly pointed head extends back to the first **gill slit** and bears a ventral **mouth** containing several rows of sharp **teeth.** The teeth, unlike those of the bony fishes and higher vertebrates, are not attached to the jaw but are embedded in the flesh. New teeth are continuously being formed and migrate forward to replace those that are lost or damaged. The head also bears two ventral **nostrils** which contain olfactory organs, the shark's primary sensory organs. The lateral **eyes** are lidless but can be covered with a fold of thin skin to protect the organ. Just caudal to the eyes is a pair of small openings, the **spiracles,** which allow water into the pharynx. The spiracles are modified gill slits that can be open or closed to regulate the amount in the respiratory system. Note that the snout extends past the mouth and that the mouth is not at the tip.

The trunk extends from the first gill slit to the **cloaca.** Five pairs of gill slits represent the external openings to the gill chamber. By inserting a probe into the spiracle and one of the gill slits you can identify the internal **pharynx.** Running longitudinally along the surface, the **lateral line** is a sensory structure which is used to detect motion in the surrounding water. The lateral line consists of a series of small, mucus-filled pores which extend back to the tail. The trunk bears the paired appendages, the **pectoral** and **pelvic fins,** as well as the **anterior** and **posterior dorsal fins.** The pectoral fins control directional changes and the pelvic fins act as stabilizers during swimming. In males, the pelvic fins

are modified to bear claspers, an accessory reproductive structure used to transfer sperm during copulation. The two dorsal fins act as rudders, stabilizing the animal from cross-currents. Both of these dorsal fins possess a strengthening anterior **spine.** The tail is heterocercal, with a larger **dorsal lobe** and a smaller **ventral lobe.**

Class Osteichthyes: *Perca*

Members of this class are called bony fish because all 21,000 species have an **ossified endoskeleton** which is more rigid than the cartilaginous skeleton of chondrichthyan fish. The skull is hardened to better protect the brain, bony plates are embedded into the skin of some species, and a hard outer sheath, the **operculum,** covers the gill chamber. Osteichthyan fish possess a **swim bladder,** an internal buoyancy organ which helps to regulate their position in the water column. In some species the swim bladder functions as a lung, absorbing oxygen across the moist membranous surface of the organ. The bodies of most osteichthyan fish are streamlined to produce efficiency of movement through the water. Most species have a **homocercal tail** with the dorsal and ventral lobes approximately equal in size. It is the tail fin which usually supplies most of the locomotive force during swimming.

Opposed to the placoid scales of chondrichthyan fish, bony fish can possess one of three types of scales. **Ganoid scales** are diamond-shaped and covered with a hard, enamel-like substance called **ganoin.** These scales impart rigidity to the fish and are commonly found in species with an armor-like exterior (gars, for example). Most osteichthyan fish possess either **ctenoid** or the

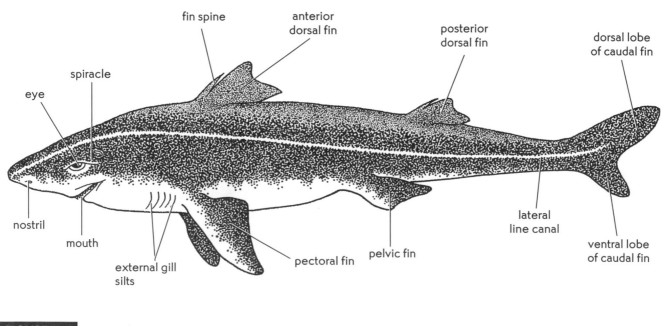

FIGURE 6.8 The dogfish shark, *Squalus acanthias.* Lateral view showing external anatomy.

more primitive **cycloid scales.** Cycloid scales are spherical and easily visible growth rings radiate from a central point. Ctenoid scales resemble cycloid scales somewhat, but the posterior edge is serrated, a factor which may reduce the amount of drag created when swimming. In general, scale size is a reflection of swimming ability. Strong, fast swimmers usually have smaller, more numerous scales whereas bottom dwellers and poor swimmers have larger, but fewer, scales.

There are numerous variations, depending upon the niche that the species occupies, on the spindle-shaped, streamlined body form of most bony fish. Generally, the more sedentary the species, the less streamlined the body form will be. Many species which inhabit coral reefs have lost the typical spindle shape and show extreme modifications in coloration and body shape. Bony fish have successfully invaded all types of aquatic environments, from freshwater streams, lakes, and ponds to brackish water estuaries to nearly all depths of marine water. There are three subclasses of bony fish: the **Dipnoi** (lungfishes); the **Crossopterygii** (coelacanths); and the **Actinopterygii** (ray-finned fishes). The first two are ancient groups with a few evolutionarily significant species.

The lungfishes are known for their ability to survive in water with little dissolved oxygen and even withstand periods of environmental desiccation by burrowing into the mud and encysting. The African lungfish, one of three living species, is known to survive for up to four years in its mud cocoon. The coelacanth is an ancient species which was thought to be extinct for 70 million years until a living specimen was caught off the coast of South Africa in 1938. It is a deep water species with strong lobe-like fins which allow it to crawl across the ocean floor for short distances. The vast majority of bony fish are ray-finned fish placed in the subclass Actinopterygii. These fish bear thin, membranous fins supported by bony rays.

Examine a variety of bony fish on demonstration. Some species which show interesting morphological adaptations are the flatfish (flounders and sole); the gars; the lungfish; the seahorse; the pufferfish; the flying fish; the eels; the viperfish; the paddlefish or spoonbill; and the sargassumfish (Figure 6.9). There are of course many more interesting species of fish. Compare the structure of some of these unusual morphologies with those of more typical bony fish, such as the freshwater yellow perch, *Perca flavescens* (Figure 6.10), and a species of bottom-dwelling catfish (Figure 6.11).

Perca

The perch body consists of three regions: the **head,** the **trunk,** and the **tail** (caudal fin). The **head** extends from the tip of the snout to the posterior edge of the **operculum,** the **trunk** ranges from the edge of the operculum to the **anus,** and the **tail** comprises the portion distal to the anus. Examine the head, and locate the terminal **mouth,** with distinct upper (the **maxillae**) and lower

jaws (the **mandibles**). The snout bears two external **nares** (or nostrils) which open to the **olfactory sacs,** sensory organs which detect levels of dissolved chemicals in the water. Fish lack an external ear, but two internal ears consist of **semicircular canals** (balancing organs) that enable the fish to maintain the proper equilibrium. The **eyes** are lateral and lidless. Behind each eye, find the **operculum,** a flap-like covering which protects the four internal **gills.**

Remove the operculum from one side, and examine the underlying gills. With a pair of forceps, pull out a single gill arch, suspend it in a small dish of water, and examine it with a dissecting microscope. Each gill bears a double row of vascular **gill filaments,** a series of **gill rakers** which protect the gills from large particles entering the gill chamber. Water enters the body through the mouth, passes through the pharynx, and penetrates the spaces between the gill filaments, called **gill slits.**

Along each side of the fish, locate the **lateral line,** a sensory organ which consists of a series of water-filled canals that detect pressure differences in the surrounding water. These pressure differences are produced by the movement of the fish itself as well as the movements of neighboring objects in the water. An **anterior dorsal fin** and a separate **posterior dorsal fin** are on the back of the perch; a **caudal fin** is located at the end of the tail; and an **anal fin** is placed on the ventral surface just anterior to the tail. Immediately in front of the anal fin are the **anal** and **urogenital openings.** The paired appendages are lateral fins located toward the anterior end. The **pectoral fins** are directly behind the opercula, and the **pelvic fins** are placed ventrally. The fins are membranous extensions of the integument supported by fin rays. The fin rays of the anterior dorsal fin are rigid and spiny; the remaining fin rays are soft and flexible. The fins are individually modified for forward movement, directional changes, stabilization, and maintaining equilibrium.

Most of the body is covered by thin, rounded ctenoid scales. Gently tug on a scale with a pair of forceps, remove it from the body, and place it in a drop of water on a microscope slide. Examine it under the low power objective of a compound microscope. Note the concentric growth rings. Compare this to the other scale types mentioned previously. As a fish grows, the size of each scale, rather than the number of scales, increases. Because growth slows with decreased temperatures, rings formed during cold seasons (fall and winter) are closer together than those formed during warmer seasons (spring and summer). Counting the number of regions of closely spaced concentric rings will provide an estimate of the number of years the specimen lived. This is not always reliable, however, depending upon the severity of the winters and the age of the fish. Scales on older fish often become worn and weathered to the point where the growth rings are obscured. Run your hand lengthwise along the body. The scales are covered by a soft protective layer of

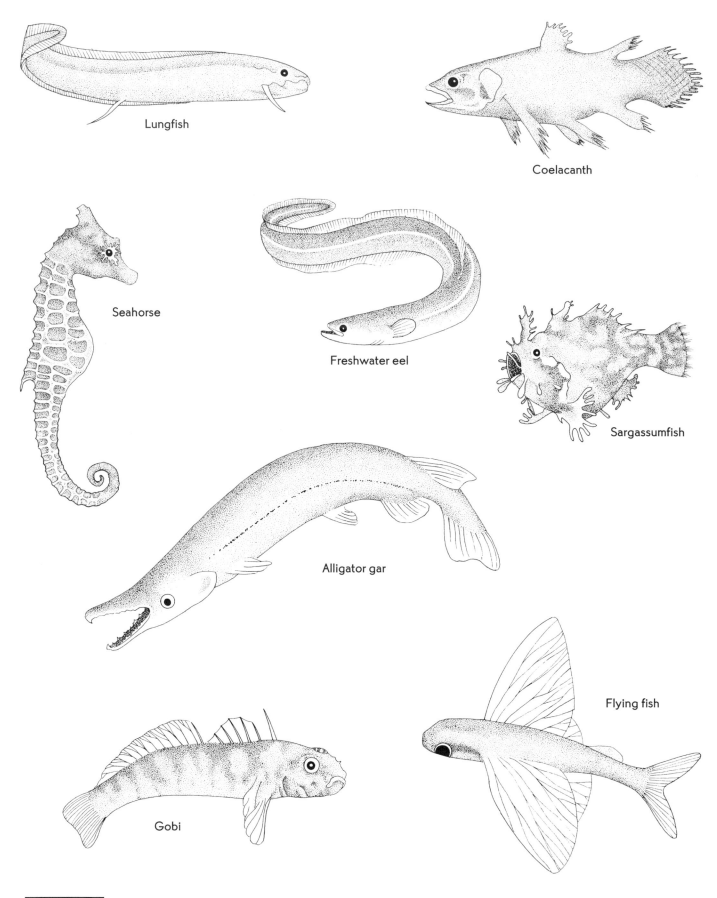

FIGURE 6.9 Various species of osteichthyan fish showing diversity of form (not to scale).

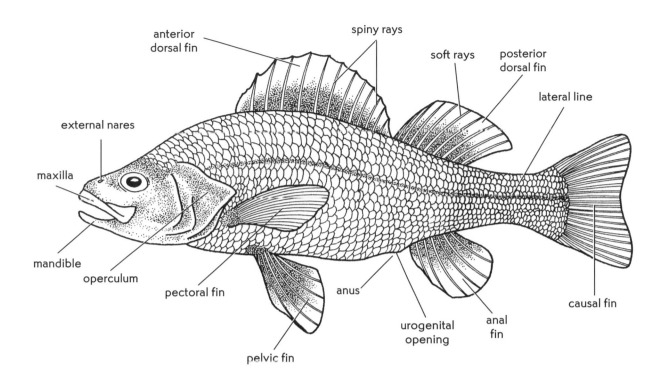

FIGURE 6.10 The yellow perch, *Perca flavescens*. Lateral view showing external anatomy.

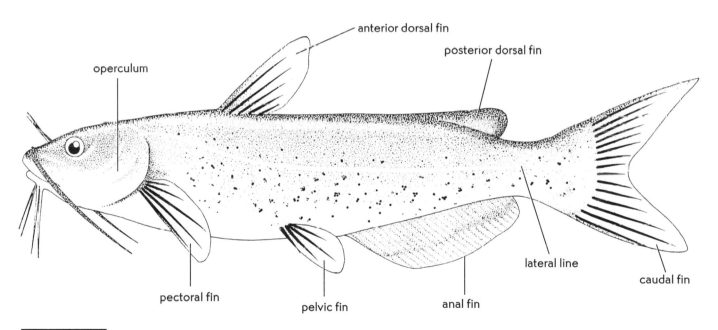

FIGURE 6.11 Lateral view of the channel catfish, *Ictalurus punctatus*.

mucus that increases efficiency of movement and helps prevent attack by parasites.

Class Amphibia

Amphibians, while perhaps not an ecologically dominant vertebrate group, occupy an important phylogenetic position. They represent the first animals to successfully colonize terrestrial environments and also form the genetic stock from which all higher vertebrates (reptiles, birds, and mammals) evolved. Approximately 300 million years ago amphibians were the dominant life form on land, but they were "quickly" replaced by their reptilian descendants within 60 million years. Ancient amphibians were often large semiaquatic animals which occupied the top of the food chain. An extinct group, the **labyrinthodonts**, were the ecological equivalents to the modern crocodilians, living along stream banks and floodplains capturing fish and other amphibians. Today, only a small remnant of a once diverse past survives. Living amphibians include the **frogs, toads, salamanders,** and burrowing **caecilians.** The modern species retain the transitional lifestyle typical of their ancestors. As such, amphibians exhibit a mixture of terrestrial and aquatic characteristics. Despite terrestrial modifications such as legs, lungs, nostrils, and sensory organs that function in air, amphibians are reproductively tied to aquatic habitats. Because their eggs lack a hard water-retaining shell, amphibians must lay their eggs in or near the water. Most species have aquatic larvae which feed for a short time before metamorphosing to a more terrestrial adult.

Amphibians possess a moist, glandular skin which is highly vascular. In fact, many species rely on **cutaneous respiration** for the majority of their gas exchange. Because the skin is not an effective barrier to moisture loss, amphibians must remain moist in order to survive. Consequently, they are found most often in either open water or damp, shady habitats with plenty of hiding places. Amphibians are some of the most ecologically sensitive animals known, and thus serve as a good marker organism to monitor environmental pollution. Unfortunately, modern amphibians are not faring well, due in part to habitat loss and human impact on the environment. They are important pharmacological organisms, producing many products useful in medicine and industry. Frogs are a food source for many human cultures, and are efficient biological controls of insect populations.

Three orders of living amphibians are known: the **Anura** (frogs and toads); the **Urodela** (salamanders); and the **Gymnophiona** (caecilians). Adult anurans characteristically lack tails and possess powerful elongate hindlimbs used in jumping. Frogs can be distinguished from toads based on their smooth skin. Toads have a bumpy, warty appearance to the skin and possess a pair of **parotid glands** just posterior to the eye and tympanic membrane (external ear). Most frogs and toads are small animals; the largest species (an African frog) reaches a weight of 3.3 kg and a length of 30 cm. The smallest frogs are less than 1 cm in length. The North American leopard frog, *Rana pipiens*, will be studied in detail as a representative vertebrate in subsequent labs (Exercises 8 and 9).

Urodeles (salamanders) are tailed amphibians as adults, with a rather unspecialized body form. The body is elongate with four equal sized limbs and a large head. The body of a typical salamander resembles the ancestral form a generalized tetrapod, a feature presumably retained from the labyrinthodonts. Most salamanders are terrestrial animals which hide about the leaf litter or under rocks and logs in damp areas. A few families have retained, as adults, the highly aquatic habits of their larvae. These aquatic salamanders, like the nearly legless amphiuma and the mudpuppies, may be mistaken for eels. Examine a few salamanders and compare body forms. Using a species of *Ambystoma*, note the external anatomy (Figure 6.12). The caecilians are a poorly known group of limbless, burrowing amphibians. At first glance they may resemble a large annelid worm due to external folds in the skin which resemble segments. Most caecilians live in the tropics where they burrow in the forest floor searching for small invertebrate prey. A few species are aquatic and live in ponds and streams. The eyes of caecilians are greatly reduced and barely functional. Replacing the eyes as a primary sensory structure are small tentacles on the snout.

Class Reptilia

The Reptilia includes animals which are morphologically similar but do not represent a natural evolutionary group. This situation arises because some reptiles, the extinct dinosaurs and living crocodilians, are more closely related to birds than to other reptiles (turtles, lizards, and snakes). To group these animals in a way which is taxonomically consistent with evolutionary patterns would destroy what was once thought to be a cohesive, homogenous taxon, the Reptilia. Because reptiles possess a suite of similar characteristics, most scientists have opted to keep the Reptilia a viable taxonomic category. Regardless, the group remains an interesting and significant assemblage of scaly tetrapods, one which is successfully adapted to dry environments. The 7,000 species of modern reptiles are placed into four major orders: the **Squamata** (lizards and snakes); the **Chelonia** (turtles); the **Crocodilia** (alligators and crocodiles); and the **Rhyncocephalia** (the tuatara, a lizard-like species restricted to some small islands off New Zealand).

A primary feature of all reptiles is the presence of dry, cornified **scales** which cover the body. The scales form an impervious barrier which retards moisture loss from the body, thus allowing reptiles to thrive in areas which are uninhabitable to amphibians. The scales come in a variety of forms and can be expanded to form shield-like plates or reinforced with dermal ossicles. When a reptile grows it must produce a new set of scales underneath the current set, thus all reptiles shed their skin periodically. Some species shed the entire skin in a single piece (snakes), while others shed small sections of skin (turtles and lizards). To minimize the energy lost when the skin is shed, some reptiles, particularly lizards, eat the skin following a molt. Another important feature of reptiles is the ability to produce a water-resistant, hard-shelled egg. This **cleidoic egg** is a great advancement over the soft unshelled eggs of amphibians. In addition, the development of an efficient **kidney** has maximized water retention and allowed reptiles to inhabit terrestrial, freshwater, and even marine environments.

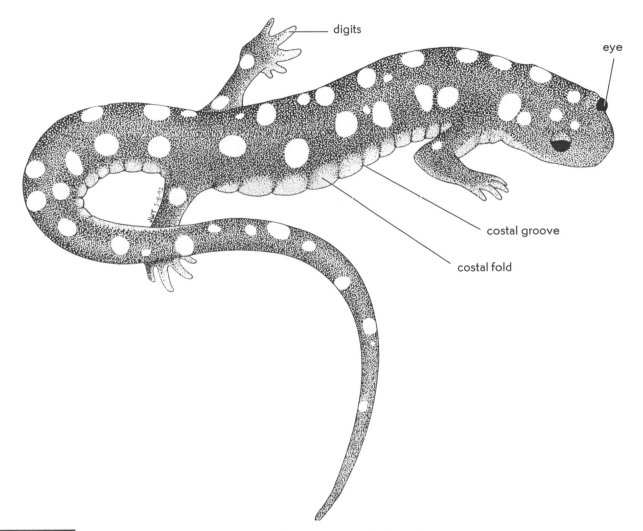

digits

eye

costal groove

costal fold

FIGURE 6.12 External anatomy of *Ambystoma maculatum*, the spotted salamander.

Reptiles are common animals across the earth. Their ecologic peak was reached in the Mesozoic era, when they dominated the earth for 165 million years. Today, reptile distribution is primarily limited by temperature; the warmer the climate the more species are present. This is because living reptiles are **ectothermic poikilotherms** (i.e., they have variable body temperatures which rely on external heat sources). Reptiles cannot generate enough internal heat through metabolism to support their physiological needs, thus, they must absorb heat through their integument from an external source. They can regulate the amount of heat being absorbed by modifying their behavior; changing the angle at which the sun's rays strike the body is one method; being active at different times of the day is another.

The chelonians, or turtles, are the most primitive living group of reptiles. They are protected by bony extensions of the rib cage which form an external armor-like covering, the **shell.** The shell is comprised of a dorsal **carapace** and a ventral **plastron** which are connected on each side by a lateral bridge. The shell is often domed and provides a spacious cavity which houses the internal organs and provides a transportable hiding place. Some species can withdraw completely into the shell; others can draw the head and legs only partially into the shell. The shell is somewhat bulky and restricts the movements of most turtles, particularly those that live on land. Aquatic turtles are more streamlined and generally better adapted for locomotion. Examine a variety of turtles on demonstration and determine whether they are terrestrial or aquatic based on the shell profile, the foot structure, and the presence or absence of webbing between the toes. Some of the more common turtles of the U.S. are the box turtle (*Terrapene carolinensis*), the snapping turtle (*Chelydra serpentina*), the pond sliders (*Pseudemys* and *Chrysemys*), and the bizarre soft-shelled turtle of the genus *Apalone* (Figure 6.13).

The largest turtles in the world are the sea turtles; some of these giants can exceed weights of 900 kg. The Galápagos tortoises (*Geochelone elephantopus*) are the largest living land tortoise and may represent some of the oldest living individual animals. A large specimen may weigh over 200 kg and live to be 150 years old. A North American freshwater species, the Alligator snapping turtle (*Macrochelys temminckii*) rivals the Galápagos tortoise in age and weight. These monstrous beasts inhabit large rivers and sluggish streams of the lower Mississippi basin where they feed on invertebrates, fish, carrion, and even some plant matter.

Snakes and lizards comprise the largest order of living reptiles, Order Squamata. Squamates represent nearly 95% of all living reptile species; presumably this great success lies in their adaptability to a variety of habitats and environmental conditions. Lizards evolved several million years before snakes at the close of the dinosaur age; their adaptive radiation produced a variety of terrestrial, aquatic, arboreal, and burrowing forms. One group of ancestral lizards, which later gave rise to the monitor lizards and the Gila monster, became highly adapted for a burrowing lifestyle. This group became elongate, reduced the external appendages, became legless, and rearranged the internal organs in response to their burrowing habits. Out of this stock the snakes evolved and refined the adaptations for a subterranean lifestyle. Subsequently, snakes

experienced an adaptive radiation of their own, and reemerged into terrestrial, aquatic, arboreal, and even marine species.

Lizards and snakes, though sharing a common ancestry, have several distinguishing characteristics: lizards possess movable eyelids, an external ear opening, and rather uniform scales which cover the entire body; snakes possess a transparent shield which covers the eye but cannot be opened or closed, they also lack an external ear and must sense vibrations through their lower jaw rather than hear airborne sounds, and they possess scales which are structurally differentiated based on location (cephalic, dorsal, ventral, subcaudal).

Lizards, with a few legless exceptions, tend to retain the generalized tetrapod body plan. They vary in size from a few centimeters in length to well over 3 m. There are a number of successful lizard families; of these the iguanas, the monitors, the geckos, the skinks, the agamids, and the true chameleons are the most conspicuous and ecologically significant. One species of monitor lizard, the Komodo dragon (*Varanus komodoensis*) is by far the largest and most dangerous species of lizard. This animal can reach lengths of 3 m, run at

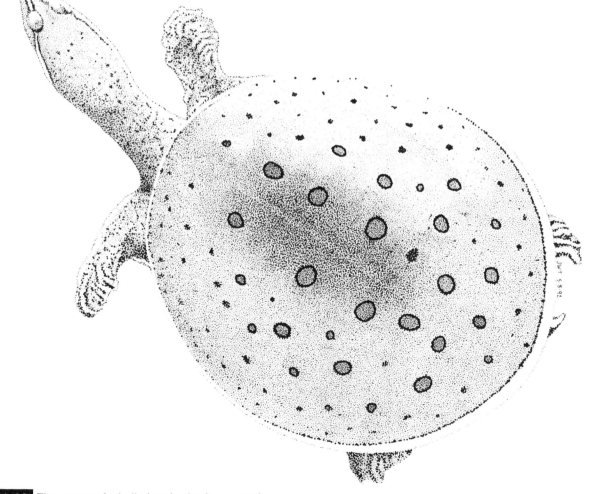

FIGURE 6.13 The spiny soft-shelled turtle, *Apalone spinifera*.

speeds nearing 50 km/h, and swallow 2 kg of meat in a single minute. These lizards generally prey on small mammals, but larger animals, such as goats, dogs, and even man can become prey. The neotropical iguanas and their old world equivalents, the agamids, are the most numerous species. Most iguanas and agamids are medium-sized (0.5 to 1.5 m) insectivorous and herbivorous lizards. Common green iguanas (*Iguana iguana*) are often sold in the U.S. as exotic pets; iguanas are a common food item in some South and Central American countries; and the marine iguana (*Amblyrhynchus cristatus*) of the Galápagos islands is unique in its marine adaptations. Not only does the marine iguana seek shelter in the sea when disturbed, but it also feeds there, swimming about the surf in search of seaweed. Examine a few lizards on demonstration and compare their structure. Note any specializations and adaptive features.

Common lizards of the U.S. include the iguanid anoles (*Anolis*, American chameleons) and horned lizards (*Phrynosoma*), the burrowing skinks (*Eumeces*) and fence lizards (*Sceloporus*), and even an introduced gecko (*Hemidactylus*, the Mediterranean gecko). Many of the geckos and skinks surprise their predators by **autotomizing** the tail when disturbed or held. The tail is suddenly cleaved from the body, leaving the confused predator holding a still-wiggling tail while the animal makes its escape. The nocturnal geckos are well known for their ability to cling to smooth surfaces, climb walls and glass, and produce loud vocalizations. The desert southwest of the U.S. harbors many interesting species. The leopard and collared lizards (*Crotaphytus*) are beautiful, swift moving species adapted to live in arid, sandy environments. The diurnal whiptail lizards (*Cnemidophorus*) are active, nervous prowlers known for their long tail and speedy escapes. The chuckwalla (*Sauromalus obesus*) is a large lizard which was utilized as a food source by Native American indians. Upon drying the hide, the lizard was often converted into a musical instrument by sewing up the hide and enclosing small rocks and pebbles and shaking back and forth. Perhaps the most infamous lizard of the U.S. is the poisonous Gila monster (*Heloderma suspectum*). This large lizard has a heavy body, massive head, and short powerful limbs. They move with a lumbering gait, but can lash out quickly when disturbed. Their venom, which is located in glands along the lower jaw, is chiefly a defensive mechanism and is ground into the prey during mastication.

As a group, snakes are limbless and most species have lost even the vestiges of the pectoral and pelvic girdles. Members of the two most primitive families of snakes, the boas and the pythons, retain portions of the pelvic girdle only. The vertebrae of snakes are short and wide, thus allowing quick lateral undulations which translate into an efficient undulatory mode of locomotion across terrestrial and aquatic environments. Snakes exhibit several types of movements generated in response to need, but most involve throwing alternating loops of the body against the ground to produce leverage. Most of the vertebrae possess lateral extensions, or ribs, which support the vertebral column, provide rigidity to the body, and increase resistance to lateral stress. In addition to the obvious loss of limbs, one of the most advantageous features of snakes is the ability to detach the mandibles (lower jaws) from each other as well as from the skull. The skull is loosely articulated and can flex asymmetrically, allowing the animal to engulf prey which is much larger than the original diameter of the mouth.

The sensory organs consist of the usual chordate assemblage of olfaction, vision, taste, and touch, but most snakes have rather poor visual acuity and rely on characteristic patterns of motion to see. As noted before, snakes are extremely sensitive to ground vibrations and may be able to sense low frequency (100–700 Hz) sound waves. Their chief sensory organ, however, is a chemosensory structure located in the roof of the mouth. **Jacobson's organs,** a pair of pit-like organs in the roof of the mouth, are richly innervated and lined with a sensitive olfactory epithelium. Snakes extend the forked tongue from the mouth, pick up scent molecules from the environment, and retract the tongue. In so doing, the tongue passes over the Jacobson's organs and sensory information is relayed to the brain. This is why active snakes repeatedly stick their tongues out, especially if disturbed or when prey items are near.

More than half of the 3,000 species of snakes belong to the family Colubridae, a diverse and widespread family of generally small snakes. This group includes many common snakes of the U.S., including the kingsnakes (*Lampropeltis*), the water snakes (*Nerodia*), the garter and ribbon snakes (*Thamnophis*), the rat snakes (*Elaphe*), the racers (*Coluber*), and the bullsnakes (*Pituophis*). These are all harmless snakes which feed on a variety of rodents, birds, frogs, fish, and other reptiles. Examine the snake specimens on demonstration and note their uniform morphology. Because snakes evolved from structurally reduced, burrowing ancestors, most species differ externally primarily in size, scalation, and coloration. Colubrids average 1–2 m in length, but many small (20–50 cm) species exist and a few colubrids (rat snakes and indigo snakes) reach lengths over 2.5 m. Colubrids and other snakes are an important biological control in regulating the populations of rodents, which introduce disease and can become agricultural pests.

Examine a specimen of the common genus *Thamnophis*, and note differences in scalation across the body. The dorsal scales are relatively uniform, the ventral scales are expanded laterally as a locomotory adaptation, and the head scales form large differentiated shields. Many species of snakes are distinguished by the size and positioning of the cranial shields. Examine Figure 6.14 and locate the identified head scales.

The largest snakes are tropical species of the boa family and the python family. The Reticulated python

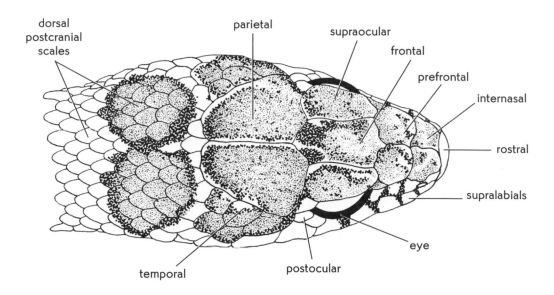

FIGURE 6.14 Head scalation of *Thamnophis sumichrasti*, a Mexican garter snake.

(*Python reticulatus*) holds the length record with some individuals nearing 10 m; most pythons, such as the Indian or Burmese python (*Python molurus*) and the African rock python (*Python sebae*) attain maximum lengths of 5–7 m. The boas (*Constrictor constrictor*) of South and Central America reach lengths of 4 m, but one semiaquatic boa, the Anaconda (*Eunectes murinus*), is known to reach lengths approaching 10 m. An unverified sighting has reported an Anaconda of 11.5 m. The gigantism of the Anaconda can probably be attributed to its aquatic lifestyle.

One family of poisonous snakes, the Viperidae, strikes fear into the hearts of men on most of the world's continents. Vipers are heavy bodied snakes with a large head and protrusible fangs. Vipers possess a hemotoxin that, when injected into the bloodstream of their victim, bursts red blood cells, destroys capillaries and the lining of blood vessels, and causes extreme swelling and pain. Old world vipers, such as the Gaboon viper (*Bitis gabonica*), the Puff adder (*Bitis arietans*), and the Horned viper (*Cerastes cornutus*) terrify the populations of Africa and the Middle East, but the actual number of yearly fatalities is fairly low.

New world pit vipers are named after a heat sensing pit located between the eye and the nostril. The organ can detect temperature fluctuations produced by warm-blooded animals nearby. In South and Central America the bushmaster (*Lachesis muta*) and the Fer-de-lance (*Bothrops atrox*) are large (3 m) species with a reputation for deadliness. Farther north, the Rattlesnakes (*Crotalus* and *Sistrurus*) are common, particularly in Mexico and the southern United States. The tail of rattlesnakes harbors a warning device, the rattle, which is a series of hollow buttons. A new button is

produced after each shed, so the more buttons the older the individual. This is not a reliable guide to reveal the age of the animal, however, because the older buttons become worn and fall off periodically. Two of the most common poisonous snakes in the United States are the semiaquatic Cottonmouth or Water moccasin (*Agkistrodon piscivorus*) and its more terrestrial relative, the Copperhead, *Agkistrodon contortrix* (Figure 6.15). These snakes probably evolved from a common Asian ancestor that migrated across the Bering land bridge several million years ago. Both are extremely common in the southeastern United States.

Although lacking in vipers, Australia has its share of poisonous snakes. In fact, Australia is one of the few places where poisonous species outnumber the harmless snakes. Members of the family Elapidae are common here, as well as in Africa, India, and Southeast Asia. Elapids possess short, stout fangs which deliver a powerful neurotoxin, causing paralysis and respiratory failure in their prey. *Notechis scutatus*, the Australian Tiger snake is arguably the most deadly terrestrial snake; the kraits, mambas, cobras (*Naja*), and neotropical coral snakes (*Micrurus*) are also common, but deadly, examples. The cobras are highly active, nervous snakes that make the ponderous heavy movements of a viper look docile. Cobras are known for their ability to imposingly raise the anterior third of the body several feet in the air just before striking. When disturbed, some species have the capability to forcibly eject venom, through a spitting action, toward the direction of the disruption. If the venom enters the body through the moist tissues of the eyes, nose, or mouth, an attacker can be invenomated.

The 21 species of crocodilians (crocodiles, alligators, caimans, and gavials) represent the last vestiges of

FIGURE 6.15 The copperhead, *Agkistrodon contortrix*, a common pit viper of the United States.

the ruling reptiles, the dinosaurs. This semiaquatic group of reptiles has remained unchanged for approximately 160 million years. Adult specimens range from 2 m to 7 m and many have no natural enemies once full size has been attained. The impingement of human culture into areas inhabited by crocodilians is chiefly responsible for their recent population decline. Some species, such as the narrow-snouted gavials (*Gavialis gangeticus*) and the American crocodile (*Crocodylus acutus*), may be facing extinction, but others like the American alligator (*Alligator mississippiensis*) and the Saltwater crocodile (*Crocodylus porosus*) appear to no longer be threatened. Crocodilians are adapted for a highly aquatic existence; their bodies are flattened dorsoventrally and their eyes and nostrils are placed on top of the head to allow the animal to see and breathe while remaining submerged. The skin of crocodiles is thick and leathery with bony plates reinforcing the scales. The head bears long, sharp teeth capable of tearing and shredding the toughest of material. The tail, used as both a locomotory organ and a weapon, is muscular and powerful. Crocodilians are adapted to slash and grab their prey with their powerful jaws and long teeth. Many crocodiles drag large prey under the water, where they drown it and leave the carcass tangled in debris for a later meal.

If available, examine the skulls of an alligator and a crocodile and note the differences. Alligators are generalized feeders, taking a variety of fish, waterfowl, mammals, and other reptiles. Crocodiles are more specialized to feed on fish, hence the snout of a crocodile is thinner in diameter than that of an alligator so that the animal can slash through the water more efficiently. The narrow snout of the gavials are an extreme example of this trend. In addition, the teeth of crocodiles, particularly the 4th tooth, are visible even when the mouth is held shut, whereas in alligators the teeth are mainly obscured when the mouth is closed.

Class Aves

Birds are among the best known and most conspicuous of the vertebrate groups (Figure 6.16). They are one of a few groups of animals that are capable of complex communication through vocalizations. Their behavior is incredibly intricate and their learning ability unequaled by any of their reptilian ancestors. Birds are the descendents of the dinosaurs, but their reptilian characteristics are limited to the presence of scales on their legs and their egg-laying habits. Most birds are capable of independent flight; this is accomplished by means of **feathers**, a crowning achievement in biological

FIGURE 6.16 Representative birds showing diversity of form.

engineering. No other animals possess feathers, although the mammalian bats can also fly independently for short periods of time.

Avian adaptations for flight include the reduction of overall body weight, and the generation of metabolic and muscular power. Features associated with weight reduction include a thin skin, light feathers, a light epidermal bill, hollow bones containing air sacs, an overall reduction in the number of skeletal elements, the lack of a urinary bladder (and the secretion of uric acid, a solid which requires no heavy liquid storage), and a high calorie diet which requires little time to digest (and hence no storage). Power adaptations include an insulated **homeothermic** body with a high metabolic rate, high blood pressure, a four-chambered heart that efficiently separates venous and arterial blood, and an efficient ventilatory system.

Fossil birds are rare due to the fragile nature of their bones, but several fossils of bird prototypes (*Archaeopteryx* and *Protoavis*) indicate the earliest birds had homodont teeth on the bill, scales on the head, and were relatively weak fliers. Living birds comprise a group of 8,600 species placed in 28 orders. Despite the fact that a few of these orders lack the capability to fly (they are adapted for swimming or bipedal running) their body architecture has been somewhat constrained by the requirements of flight. Flying birds must remain aerodynamic and have rigid weight requirements. Because of these constraints, most of the adaptive features of birds center around the bill, feet, and legs.

Examine specimens from several orders of birds (you may have to visit your local univeristy or state museum) and note the morphological differences in bill length, leg length, tail length, and foot structure of a variety of species (Figure 6.16). You can use predictive morphology to assess the ecology of the species. Webbed feet indicate the presence of an aqautic species; a serrated bill is usually associated with piscivory; short, stout bills are indicative of seed eaters, while long narrow bills are typical of insectivorous species. On an insect eater, note the stiff bristles above the bill which protect the eye from damage by struggling prey. Most of these feeding adaptations form a suite of characters; for example, woodpeckers are adapted to cling to tree bark and drill for their insect prey. Note the heavy nature of the bill and observe the large claws on the feet of a woodpecker. The tail is also unusually stiff to support the animals as it presses against the tree. Birds that are excellent fliers usually possess a wide wing span and a well developed tail. Species with a short tail but a wide wing span are usually very acrobatic and highly maneuverable.

Some common orders of North American birds are: **Order Gaviiformes** (the loons); **Order Procellariiformes** (the albatross, petrels, and shearwaters); **Order Pelecaniformes** (the pelicans, boobies, and cormorants); **Order Ciconiiformes** (the herons, egrets, storks, and spoonbills); **Order Anseriformes** (the ducks, geese, and swans); **Order Falconiformes** (the hawks, eagles, ospreys, falcons, vultures, and condors); **Order Galliformes** (the domestic chicken, turkeys, pheasants, and quail); **Order Strigiformes** (the owls); **Order Piciformes** (the woodpeckers and sapsuckers); and **Order Passeriformes** (all of the perching songbirds).

Class Mammalias

Although only consisting of 4,000 species, mammals are exceedingly diverse in size, form, and shape. Mammals range in size from the tiny 1.5 g Kitti's hog-nosed bat to the incredibly immense whales, which can reach weights exceeding 100 tons. Mammals are the most advanced group of animals; they are **endothermic homeotherms** which nourish their young with milk produced by the **mammary glands.** Mammals are characterized by possessing **hair,** a derivative of the reptilian scale. The hair forms an insulatory layer which can be thickened to form a coat or **pelage.** Mammals also possess highly differentiated **heterodont teeth** which are specialized for different mastication functions. **Incisors** and **canine** teeth form sharp cutting and slicing edges; **premolars** and **molars** have a larger surface area for grinding and chewing.

All mammals show internal fertilization and copulation. Most mammals give birth to live young (vivipary); only two rare exceptions exist, the egg-laying Duck-billed platypus and the echidna, or spiny anteater. Among the viviparous mammals, two major groups exist: the **marsupials** and the **eutherians** (or **placental** mammals). Marsupials are most prevalent in Australia because they were isolated from competition with the eutherians throughout much of their evolutionary history. Marsupials are named for their pouch-like **marsupium** that shelters the altricial developing young. This is in contrast to the reproductive strategy of the placental mammals which nurture their young in the uterus with a vascular **placenta.**

Common marsupials include the Australian kangaroos, wallabies, bandicoots, koalas, and wombats, as well as the North American oppossum, *Didelphis* (Figure 6.17). The placental mammals are extremely successful; they have exploited nearly every known environment and display a great diversity of form and structure. Some of the most numerous placentals are placed in the **Order Rodentia** (Figure 6.18). These are small active animals with an incredibly high metabolic rate, short life cycles, and great fecundity. Members of the **Order Carnivora** comprise a fascinating group of dogs, cats, bears, raccoons, skunks, weasels, and otters. All of these species are either meat-eaters or omnivores. Most are well-adapted for running and have an acute sense of smell.

The hoofed mammals, or **ungulates,** are divided into the orders **Perissodactyla** and **Artiodactyla** based on the structure of their digits. Both orders are herbivorous

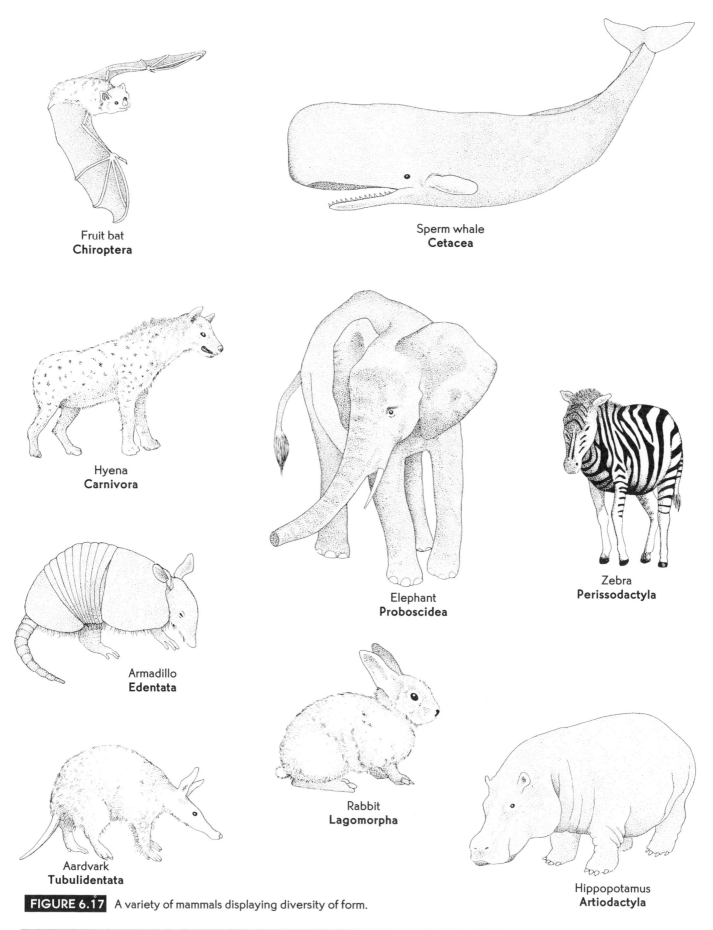

Fruit bat
Chiroptera

Sperm whale
Cetacea

Hyena
Carnivora

Elephant
Proboscidea

Zebra
Perissodactyla

Armadillo
Edentata

Aardvark
Tubulidentata

Rabbit
Lagomorpha

Hippopotamus
Artiodactyla

FIGURE 6.17 A variety of mammals displaying diversity of form.

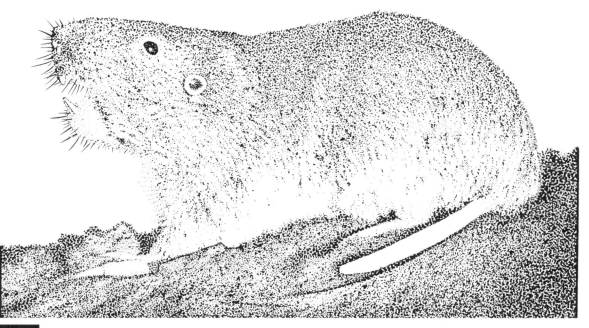

FIGURE 6.18 The pocket gopher *Geomys*, a burrowing rodent. Note the reduced ears, small eyes, and large incisors.

and have elongate limbs adapted for running. The main axis of a perissodactyl's leg passes through the third digit only, while the central axis of an artiodactyl passes through two digits. The artiodactyls are more diverse than the perissodactyls; artiodactyls include the hippopotamuses, antelope, pigs, cows, sheep, deer, and camels. The perissodactyls are limited to the horses, rhinoceroses, and the tapirs.

Animals in the **Order Lagomorpha** superficially resemble rodents and may be related to them through a recent common ancestor. Similar to the rodents the pikas, hares, and rabbits have two pairs of constantly growing incisors. Lagomorphs, however, also possess a second pair of incisors in the upper jaw.

Mammals of the **Order Insectivora** are often small terrestrial or semiaquatic animals with tiny eyes and long, tapered snouts. Shrews, moles, and the European hedgehog are common examples. The teeth of insectivores typically bear sharp cusps which they use during their nocturnal feedings on small insects.

The bats, **Order Chiroptera**, are perhaps the most interesting and unique group of mammals. While some mammals can glide for a short distance on air currents, only the bats can generate true flight among mammals. Most species are insectivorous and nocturnal, but many herbivorous and frugivorous species are known, and some bats capture fish, frogs, birds, and even other bats. Bats are generally small and follow the rodent physiological plan. They are secretive animals, hiding in crevices and caves during daylight hours and actively foraging at night. Their primary sensory mechanism is **echolocation,** a form of radar that enables the bat to sense oncoming objects in the absence of light. Most bats are harmless to man, but the notorius *Desmodus* (vampire bats) have generated a horrific reputation by feeding on the blood of large mammals. Vampire bats occasionally attack humans and can transmit rabies. Their razor-sharp incisors and canines pierce the skin of their victim and they lick up the blood as it flows out. As a mechanism to reduce their flying weight after engorging on a meal, they often urinate on their victim while feeding.

The most advanced animals in the world belong to the **Order Primates.** Most species are arboreal and occur in tropical and subtropical Africa, Southeast Asia, and South America. Many of the primates have an extremely well-developed nervous system and a highly adaptive brain. Monkeys, the great apes, marmosets, lemurs, lorises, and human beings are some important primates.

Perhaps the group of mammals which have undergone the most radical adaptations to their environment are the marine mammals of the **Order Cetacea.** These whales, dolphins, and porpoises are streamlined animals which possess sparse amounts of hair as a swimming adaptation. All cetaceans breathe oxygen through lungs, but cetaceans have a suite of mechanisms, both behavioral and structural, which enable them to remain submerged for long periods of time. Cetacean communication is intricate and incredibly complex; whistles, clicks, grunts, and long groans are all used in combination to signify a message. Cetaceans generally interact socially and have an excellent learning capacity.

Checklist of Suggested Demonstrations

THE HEMICHORDATES AND CHORDATES

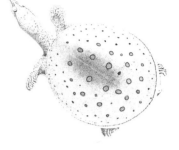

_____ 1. *Saccoglossus* (or *Balanoglossus* or *Dolichoglossus*) preserved specimen.

_____ 2. *Ciona* preserved specimen.

_____ 3. *Molgula* preserved specimen.

_____ 4. Urochordate tadpole larva whole-mount slide.

_____ 5. *Amphioxus* whole-mount slide.

_____ 6. *Amphioxus* cross-section slide.

_____ 7. *Petromyzon* preserved specimen.

_____ 8. *Squalus* preserved specimen.

_____ 9. Preserved specimens of various skates and rays.

_____ 10. *Perca* preserved specimen.

_____ 11. Various teleost preserved specimens.

_____ 12. Frog life cycle diagram.

_____ 13. *Amphiuma* preserved specimen.

_____ 14. *Xenopus* preserved specimen.

_____ 15. *Rana* preserved specimen.

_____ 16. Preserved ranid tadpoles.

_____ 17. *Bufo* preserved specimen.

_____ 18. Preserved *Bufo* tadpoles.

_____ 19. *Apalone* preserved specimen and other common turtles.

_____ 20. Various preserved lizard specimens.

_____ 21. *Agkistrodon contortrix* (copperhead) preserved specimen.

_____ 22. *Agkistrodon piscivorus* (cottonmouth) preserved specimen.

_____ 23. *Crotalus* (rattlesnake) preserved specimen.

_____ 24. *Lampropeltis* (kingsnake) preserved specimen.

_____ 25. *Thamnophis* (garter and ribbon snake) preserved specimens.

_____ 26. *Elaphe* (rat snake) preserved specimen.

_____ 27. *Nerodia* (water snake) preserved specimen.

_____ 28. Various bird study skins or supplement with a field trip to a local museum.

_____ 29. Various mammal study skins or supplement with a field trip to a local museum.

Hemichordate and Chordate Notes and Drawings

Hemichordate and Chordate Notes and Drawings

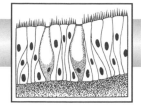

Tissue Structure and Function

In multicellular animals, cells are arranged into highly organized groups or layers. These **tissue** layers consist of aggregations of similar cells and cell products which unite to increase functional efficiency. Tissues are specialized cellular groups that exhibit the typical protoplasmic properties: contractility, conductivity, irritability, absorption, and excretion are common functions. The study of tissues, particularly their structure and arrangement, is an intricate science called **histology**. Tissues are interactive; they combine to form complex **organs** which are specialized to perform certain functions, such as circulation, reproduction, respiration, locomotion, excretion, etc. In advanced animals, the organs work as a unit, forming **organ systems.**

All animals, despite great differences in structural complexity, are composed of only four major tissue types. These basic tissues are: **epithelial tissue, connective tissue, muscle tissue,** and **nervous tissue.** Each performs a specific function and works in conjunction with other tissues to produce a functional organ.

EPITHELIAL TISSUE

An **epithelium** is a thin, sheet-like layer of cells that covers an external surface or lines an internal surface. This surface may be quite extensive (such as the outer layer of the skin and lining of the digestive tract) or quite limited (glands and microscopic tubules). The primary function of an epithelium is protection, but some specialized epithelia absorb, secrete, or excrete materials. One surface of an epithelium rests on a bed of connective tissue; the opposing surface is free and exposed to body fluids or air. Epithelial tissue lacks a blood supply; it is thin enough to allow nutrients and gases to enter through diffusion from the underlying **basement membrane,** a layer of intercellular substance between the epithelium and the connective tissue. Unlike the other tissues, epithelia are derived from all three of the embryonic layers: ectoderm, mesoderm, and endoderm.

Epithelial tissue is classified according to both the number of layers of cells and the shape of the individual cells involved. If an epithelium consists of only a single layer of cells it is termed **simple epithelium;** if two or more layers of cells are stacked on each other it is called **stratified epithelium** (Figure 7.1). The shape of the cells varies based on functional need. Some cells are thin and flat in cross-section (**squamous**), while others are cube-shaped (**cuboidal**), and still others are longer than they are wide, or column-shaped (**columnar**). Examine prepared slides of epithelial tissues and compare them to the appropriate micrographs (see figures).

Simple Epithelium

Simple epithelium is made up of a single layer of epithelial cells, all of which extend to the basement membrane. It occurs in areas of the body which experience little wear and tear or where diffusion or absorption occurs through a membrane.

Simple Squamous Epithelium

Simple squamous epithelium is composed of thin, flat cells with centrally placed nuclei (Figure 7.2). The cells appear irregular in surface view; in cross-section they appear extremely thin but bulge slightly where the nucleus is located.

Simple squamous epithelium is specialized for diffusion. It lines the walls of blood vessels and the heart, and makes up the entire capillary network. The peritoneal lining of the body cavity, the lining of the mouth and vaginal cavities, as well as Bowman's capsules of the kidney, all consist of simple squamous epithelial cells.

Simple Cuboidal Epithelium

Simple cuboidal cells are square-shaped cells that provide a little more durability than squamous cells (Figure 7.3). It occurs in areas where transport is important, such as in the kidney tubules, salivary glands, mucous glands, and thyroid follicles.

Simple Columnar Epithelium

Simple columnar cells are taller than they are wide and are often tightly packed (Figure 7.4). In vertical section they appear as a row of rectangles, with the nuclei characteristically located in the lower half of the cell.

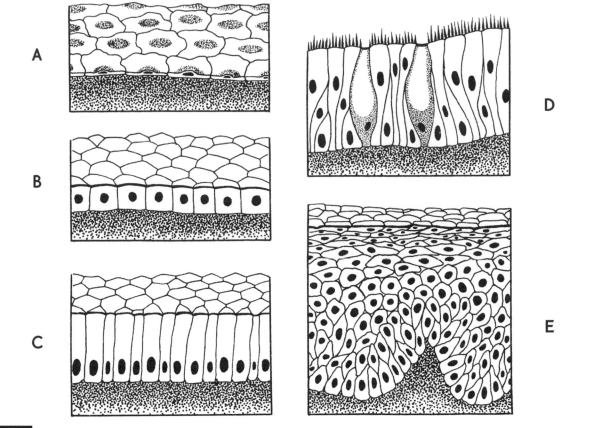

FIGURE 7.1 Various types of epithelial tissue. A) simple squamous; B) simple cuboidal; C) simple columnar; D) pseudostratified ciliated columnar; E) stratified squamous.

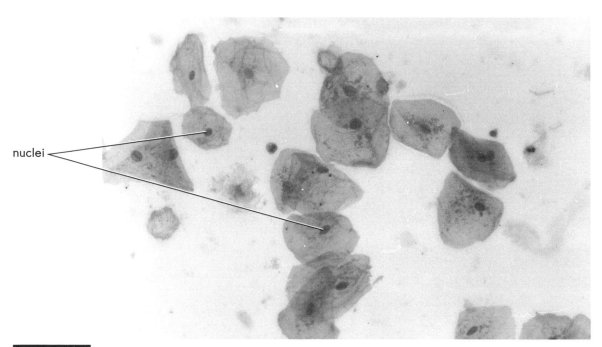

nuclei

FIGURE 7.2 Simple squamous epithelial cells, surface view.

Columnar cells occur in areas subject to high stress, but where absorption and secretion are still necessary, such as the digestive tract. The free surface of columnar cells is often ciliated, but it may be covered with a nonliving cuticle. Some specialized columnar epithelial cells are found in the lining of the digestive tract. These **goblet cells** represent single-celled glands designed to produce and secrete mucus to aid in digestion (Figure 7.5).

Pseudostratified Columnar Epithelium

Despite appearing layered, pseudostratified columnar epithelium is actually a simple epithelium with all the cells resting on the basement membrane. The cells appear stratified because they have varying heights and their nuclei are placed at different levels (Figure 7.6). Pseudostratified columnar cells are usually ciliated on the free surface and are important mucus-secreting cells

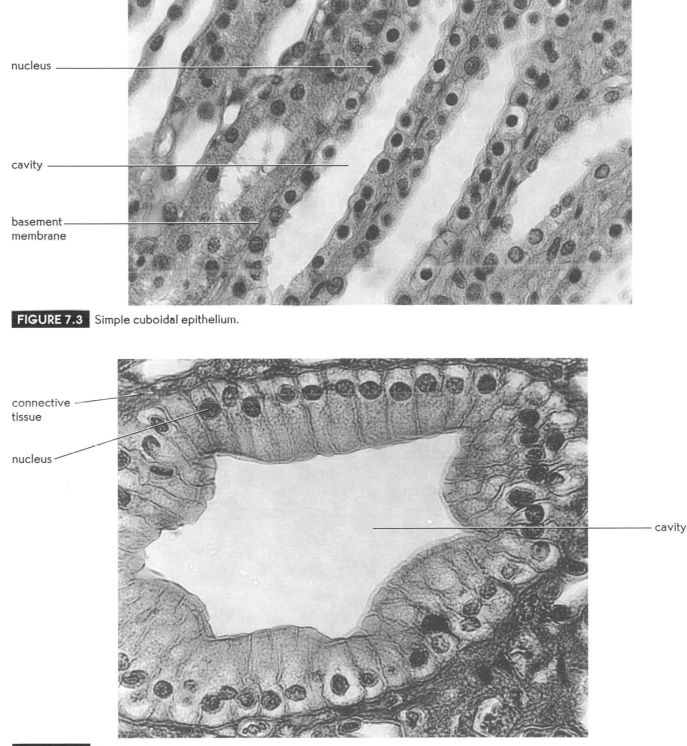

FIGURE 7.3 Simple cuboidal epithelium.

nucleus

cavity

basement membrane

connective tissue

nucleus

cavity

FIGURE 7.4 Simple columnar epithelium.

of the pharynx, trachea, and bronchi. They also comprise the urethral lining.

Stratified Epithelium

Stratified epithelium is composed of two or more layers of cells. It is found in areas where great wear and tear is experienced. Only the lower layer of cells is actively reproducing; the upper layer consists of old, dead cells that are continually being sloughed off. It is not useful for absorption or secretion due to the thickness of the tissue.

Stratified epithelium is named according to the shape of the cells at the free surface; **stratified squamous** cells comprise the outer layer of the skin, the lining of the mouth, esophagus, anus, vagina, and the cornea (Figure 7.7); **stratified cuboidal** cells line the ducts of sweat glands and the testis tubules; and **stratified columnar** lines parts of the pharynx, larynx, urethra, and salivary ducts (Figure 7.8).

lumen

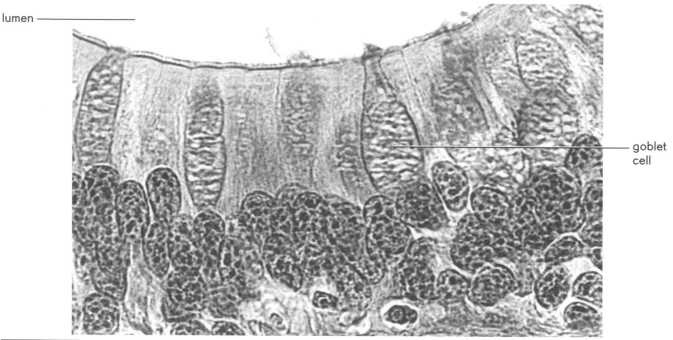

goblet cell

FIGURE 7.5 Simple columnar epithelial cells of the intestinal lining showing goblet cells.

cilia

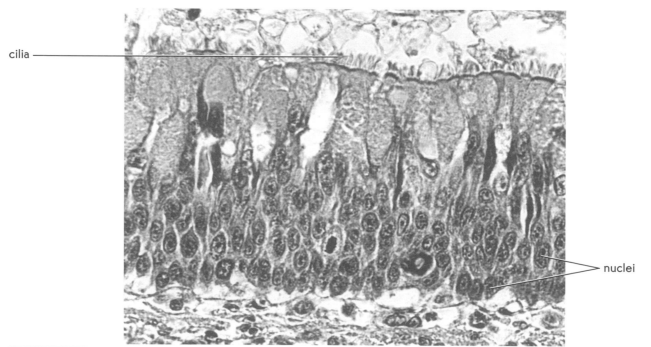

nuclei

FIGURE 7.6 Pseudostratified columnar epithelium. All cells rest on a basement membrane despite appearing stratified.

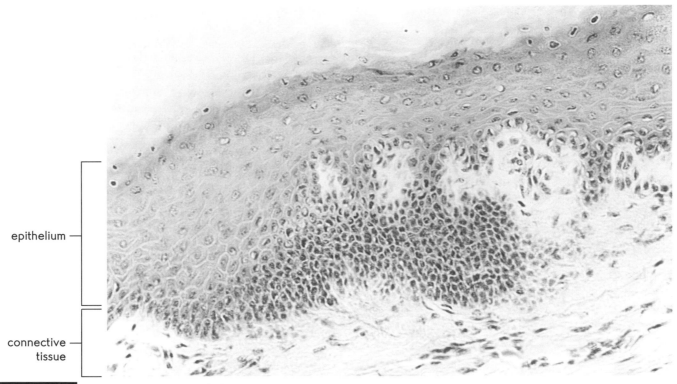

epithelium

connective tissue

FIGURE 7.7 Stratified squamous epithelial tissue.

2nd layer   1st layer

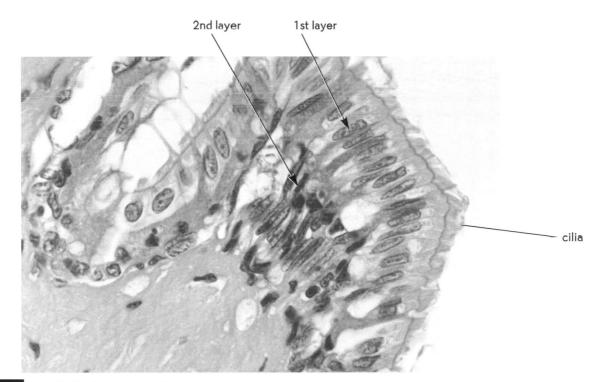

cilia

FIGURE 7.8 Stratified columnar epithelium.

CONNECTIVE TISSUE

Connective tissue, like its name implies, provides structural and metabolic support for the body by holding other tissues and body parts together. Connective tissue is very diverse in form and is often the most common tissue of an animal. All connective tissues are composed of cells embedded in an **acellular matrix,** or ground substance, reinforced by proteinaceous **fibers.** Different types of cells and fibers, along with varying densities of each, combine to produce specific connective tissues. These include **loose connective tissue** (areolar and adipose), which forms a stretchy "fabric" that surrounds specialized cells, underlies epithelial tissues, and contributes in several important ways to many other structures in the body; and **dense connective tissue** (ligament, tendons, cartilage, and bone), which supports the body structurally. Because blood cells are derived from bone marrow and flow in a fluid acellular matrix, **vascular tissue** is considered to be a highly specialized form of connective tissue.

The function of specific connective tissues depends largely on the type of fibers involved, the density of cells present, and the amount of ground substance. For example, adipose (fat) tissue is composed mainly of cells, tendons and ligaments are mostly fibers, and some cartilage is largely ground substance. There are three kinds of fibers which differ in physical properties: **collagenous fibers, elastic fibers,** and **reticular fibers.** Collagenous fibers occur in thick, wavy bundles of thinner fibrils. They are very strong and resistant to stretching. Elastic fibers appear as thin, straight threads that diverge from each other in different directions. They may be stretched to 150% or more of their resting length, but will spring back to their original length upon release. Reticular fibers are inelastic and consist of bundles of small fibrils which branch and reunite to form a network. The arrangement of these fibers is related to function. The fibers of **fasciae** (fibrous tissue that surrounds organs and holds muscles into bundles) are interlaced, predominantly collagenous, and arranged in parallel fashion. This interlaced arrangement provides strong resistance to stress from all directions. The vocal cords of vertebrates primarily consist of elastic, parallel fibers. Parallel fibers are designed to withstand tension from a particular direction. Examine slides of loose and dense connective tissue. Then compare these with a slide of human blood and note the variety of arrangements of cells, fibers, and matrix.

Loose Connective Tissue

Areolar Tissue

The most widespread of all the tissues, areolar connective tissue is a loose fibroelastic tissue specialized to fasten down the skin, membranes, vessels, and nerves, as well as binding muscles and organs together (Figure 7.9). In a fresh specimen it appears whitish, translucent, flexible, and stretchy. It consists of a clear, gelatinous matrix, some dispersed cells, and all three fiber types.

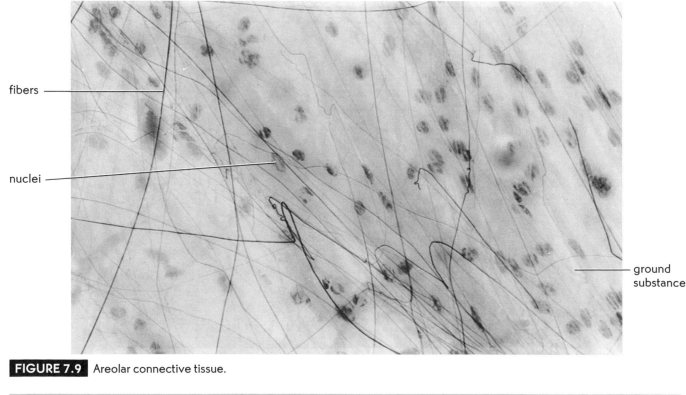

fibers

nuclei

ground substance

FIGURE 7.9 Areolar connective tissue.

Adipose (Fat)

Adipose cells are specialized for fat storage and do not require the production of fibers or a ground substance (Figure 7.10). The cells appear clear, empty, and bubble-like, but this appearance is due, in part, to the preparation of the slide. Due to the hydrophobic actions of lipids suspended in the watery cytoplasm, the nucleus and cytoplasm have been pushed to one side of the cell to accommodate the globule of fat that fills the cell. When the slide is prepared, the lipids are dissolved in alcohol, causing the cell to appear empty.

Dense Connective Tissue

Cartilage

Several forms of cartilage exist in animals, the most common of which is **hyaline cartilage** (Figure 7.11). It occurs on the ends of long bones, where it forms a soft pad to help reduce the friction generated by the joint; it's also found in the nose and a few other places, such as the trachea, where it forms supportive rings. The ground substance is firm but flexible and scattered through it are small vacuities, called **lacunae,** each of which contains at least one cell. The cells, called **chondrocytes,** secrete the ground substance, but do not produce fibers. Two or more chondrocytes in a single lacuna indicate recent mitotic activity. Other forms of cartilage include **fibrocartilage** and **elastic cartilage.** Fibrocartilage contains many inelastic fibers and occurs in the intervertebral discs, tendons, and ligaments. Elastic cartilage, composed of many elastic fibers embedded in the ground substance, is found supporting the external ear, epiglottis, and the larynx.

Bone

Bone is an extremely important tissue of vertebrates. Not only does it supply mechanical support to the body, but it is a very active tissue metabolically. Bone protects the vital organs, forms most of the necessary blood cells, and functions in locomotion. Bone is also an important calcium storage area with the capability to dole out calcium ions when needed. The matrix of bone is packed with calcium and phosphate salts secreted by **osteoblasts** or bone-forming cells.

The density of bone varies according to function. Long bones that run with the axis of the body consist of an inner core of gelatinous **marrow** (which produces red and white blood cells), and an outer shaft of **dense** (or **compact**) **bone.** The ends of the bone consist mainly of **spongy** (or **cancellous**) **bone.** Compact or dense bone consists of a series of tightly packed calcified cells termed **osteons** (Figure 7.12). Spongy bone is made of a loosely interlaced network of bony tissue interspersed with marrow. A thin membrane, the **periosteum,** covers the surface of the entire bone. Flat bones, such as those of the skull, ribs, or dermal elements of some lower vertebrates, possess a thin inner region of spongy bone surrounded by layers of dense bone.

Though hardened and rigid, bone is living tissue and must be penetrated by nerves and blood vessels. These can easily penetrate through the loose network of spongy bone but they must be routed through specific channels in dense bone. Examine a slide of dense bone and note the arrangement of the osteons, the interstitial bone between osteons, and the structures visible in a single osteon (Figure 7.13). The central **osteon canal** (or **haversian canal**) serves as a passageway for blood

FIGURE 7.10 Adipose tissue. Lipid (fat) droplets accumulate into globules in each cell, forcing the cell cytoplasm and nucleus to the edge of the cell.

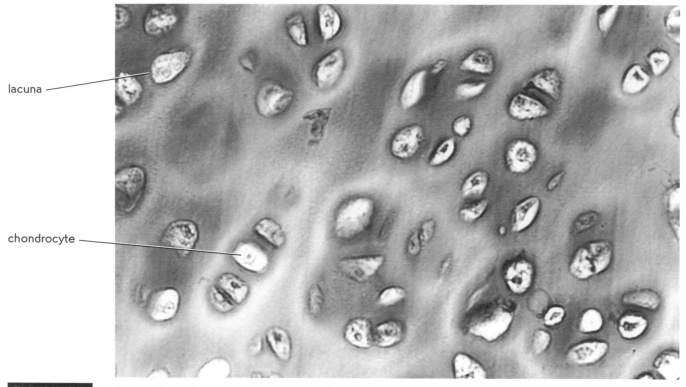

FIGURE 7.11 Hyaline cartilage showing lacunae and chondrocytes.

lacuna

chondrocyte

vessels, nerves, and lymphatic vessels. Arranged concentrically around the osteon canal are layers of bony tissue, the **lamellae.** Between the lamellar layers are darkly-stained **lacunae** that house osteocytes and act as a nutrient pool. The lacunae communicate with each other and the osteon canal by means of small fissures in the lamellae termed **canaliculi.**

Vascular Tissue

Vascular tissue is responsible for the transportation of a variety of nutrients, hormones, and gaseous oxygen to nearly all areas of the body. The carriers of these substances are blood and lymph, both specialized connective tissues with a variety of cell types. The cells, or corpuscles, flow in a fluid matrix termed **plasma.** The

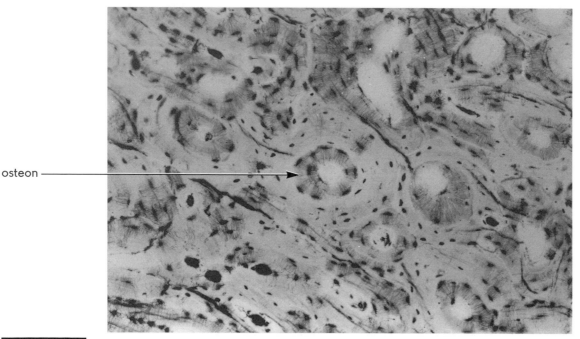

osteon

FIGURE 7.12 Cross-section through compact bone showing the arrangement of osteons.

Introductory Zoology Laboratory Guide

FIGURE 7.13 Enlarged view of a single osteon. Bone-secreting cells (osteocytes) reside in the lacunae. Osteocytes maintain the dynamic state of bone, releasing or depositing calcium as required for calcium homeostasis.

plasma is held within a system of vessels for quick delivery to specific tissues. Blood is transferred directly from the heart to the organ systems and lymph and tissue fluids are filtered from the blood to bathe all body cells. Examine a slide of human blood and note the abundance of different cell types (Figure 7.14). The most common cells are small, biconcave **erythrocytes** (red blood cells). White blood cells, **leukocytes,** occur less frequently in the blood and are much larger. Look closely to observe some tiny irregular-shaped discs in the plasma. These are **platelets,** agents responsible for clotting.

Erythrocytes

In mammals, including man, erythrocytes are round, biconcave discs that characteristically lack nuclei. They are thin and possess a lot of surface area relative to volume. Most other vertebrates possess oval-shaped erythrocytes with large granular nuclei. Vertebrate red corpuscles are packed with the respiratory pigment **hemoglobin,** which functions to transport the respiratory gases (O_2 and CO_2) to and from the lungs and body tissues. Erythrocytes are rare in invertebrates; instead, the respiratory pigments (mainly **hemocyanin**) bind into giant molecules that circulate in the plasma.

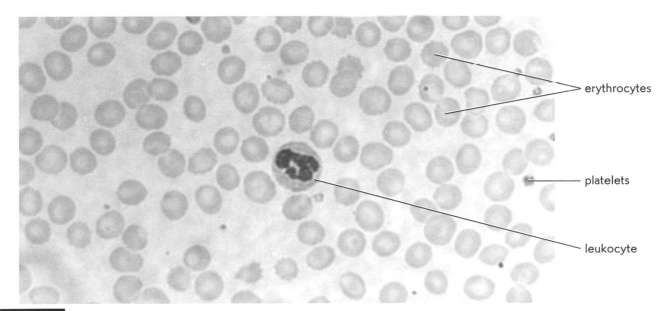

FIGURE 7.14 Human blood showing different cell types. Mammalian red blood cells (erythrocytes) lack nuclei, while white blood cells (leukocytes) have multi-lobed nuclei. Platelets are clotting elements.

Leukocytes

Vertebrate blood contains several different types of white blood cells. All contain large, darkly-stained nuclei. In human blood, leukocytes are larger than erythrocytes, but in other vertebrates they are generally smaller. Leukocytes are a primary defense against invading organisms and foreign proteins. Through phagocytosis, leukocytes can surround and engulf particulate matter, creating a fluid (**pus**) consisting of dead leukocytes and potentially harmful material. Common types of white blood cells in man are: **monocytes, neutrophils, lymphocytes, basophils,** and **eosinophils.** Each of these can be differentiated based on nuclear structure. Monocytes and neutrophils are primarily phagocytic cells, and lymphocytes are active in immune responses. Leukocytes form a wandering defense system and perform many of their protective functions outside the blood vessels in connective tissue.

Platelets

Platelets are tiny, disc-shaped clotting elements which occur only in mammalian blood. Most lower vertebrates have spindle-shaped clotting cells called **thrombocytes,** and a few invertebrates possess platelet-like cells.

MUSCLE TISSUE

Muscle tissue is composed of elongate fibers specialized for contraction. The fibers are bound in a framework of connective tissue that culminates in attachment to the skeleton or skin. The contraction of muscle fibers is produced by a rearrangement of their protein molecules and the energy is supplied and replenished by metabolizing food products. Three types of muscle exist, each of which is adapted to produce slightly different types of contraction. **Smooth muscle** is designed to perform long, slow sustained contractions; **skeletal muscle** performs quick, rapid contractions; and **cardiac muscle** is adapted to perform long term, highly synchronized rhythmic contractions. Each has distinguishing structural characteristics, as noted below.

Smooth Muscle

Smooth muscle is termed **nonstriated, visceral,** or **involuntary muscle** (Figure 7.15). It is the simplest of the three muscle tissues. Each cell is long, spindle-shaped, and uninucleate; it is involuntary in action and controlled by the **autonomic nervous system.** Smooth muscle is located in the digestive tract, uterus, and other places where long, powerful contractions are necessary.

The muscle fibers consist of bundles of slender myofibrils, which in smooth muscle do not show cross-striations, as seen in skeletal and cardiac muscle.

Skeletal Muscle

Skeletal muscle is often referred to as **striated** or **voluntary muscle** because each myofibril contains alternating dark and light bands which impart a striped or striated appearance and it is under conscious cerebral control of the **central nervous system** (Figure 7.16). Skeletal muscle fibers contract more quickly than smooth muscle, but tend to fatigue more easily.

Skeletal muscle fibers (1–40 mm) are much longer than smooth muscle fibers (0.2–0.5 mm) and are multinucleate. Each fiber is made up of many myofibrils, all enclosed in a tough outer membrane termed the **sarcolemma.** The muscle fibers are grouped into bundles, and the bundles are grouped into functional skeletal muscles. Fibers, bundles, and muscles are bound by an

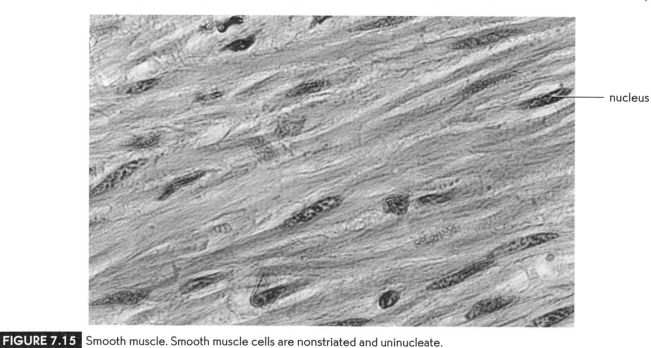

nucleus

FIGURE 7.15 Smooth muscle. Smooth muscle cells are nonstriated and uninucleate.

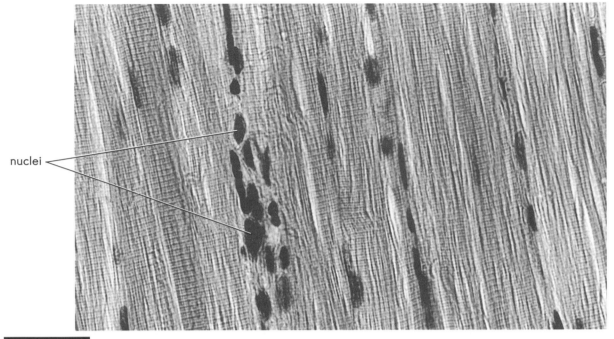

nuclei

FIGURE 7.16 Skeletal muscle. The cells are characterized by long, unbranched fibers with flattened nuclei.

outer sheath of loose connective tissue termed **fascia**. On both ends of a skeletal muscle the connective tissue surrounding the muscle fuses with the fibrous **tendon** that anchors the muscle to the bone. When the muscle contracts, it causes movement of one bone relative to the other. Observe the slide on demonstration. Note the cross-striations, the elongate multinucleate fibers, and the congregation of nuclei along the sides of the fibers.

Cardiac Muscle

Cardiac muscle is a highly specialized and regionalized muscle type; it is adapted to perform millions of synchronous contractions throughout the lifespan of the animal (Figure 7.17). It appears similar to skeletal muscle in structure but functions more like smooth muscle. It is **involuntary** and restricted to the walls of the heart

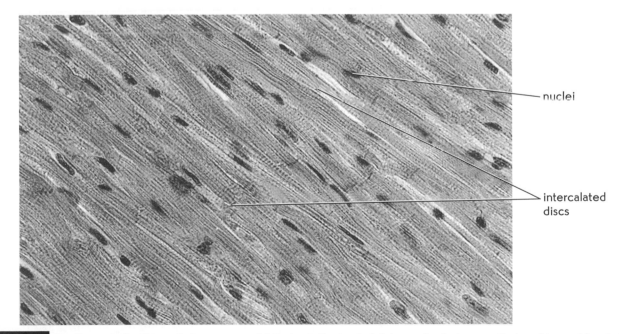

nuclei

intercalated discs

FIGURE 7.17 Cardiac muscle. The cells have cross-striations similar to skeletal muscles but are united end to end by darkly-stained intercalated discs.

and surrounding blood vessels. The fibers are arranged in branching columns interconnected into a fine network. The myofibrils are transversely **striated,** as in skeletal muscle, but the nuclei are centrally located and placed one per fiber, as in smooth muscle. Look very closely at a prepared slide. The ends of the cells are joined by dark bands called **intercalated discs.** These are actually plasma membranes specialized for conducting electrical impulses between cells, one characteristic which facilitates the rhythmic beating of the heart.

Nervous Tissue

Nervous Systems

The entire nervous system consists of a complex network of several interactive systems. The **sensory system** receives information regarding external environmental factors and internal metabolic functions. It transforms this information into nerve impulses that are conducted along **sensory (afferent) nerves.** To process and assess this information, most animals have a **central nervous system (CNS),** consisting of a brain, or ganglia, and one or more nerve cords. The central nervous system then generates motor impulses that travel through the **motor system** along **motor (efferent) nerves.** Motor control of the skeletal musculature is delegated to the **somatic (voluntary) nervous system;** while control of the smooth, or visceral, muscle and glandular activity is regulated to the **autonomic nervous system.** Sensory nerves that provide information to the central nervous system and motor nerves that control muscle and organ activity are collectively referred to as the **peripheral nervous system.**

Neurons

Nervous tissue is perhaps the most highly specialized of the four basic tissue groups. It is adapted for the reception of environmental stimuli and for the conduction of nervous impulses. Nervous tissue is composed of specialized cells called **neurons.** Neurons vary in their form and function, but all consist of a nucleated **nerve cell body** and many fibrous extensions called **nerve cell processes** or **nerve fibers.** The processes that carry impulses toward the nerve cell are called **dendrites;** those that carry impulses away from the cell are termed **axons.** Dendrites are usually short and highly branched, and outnumber the longer axons. In some large vertebrates, a single axon may extend for several feet, bringing motor impulses from the spinal cord to the end of the appendages.

Axons of the peripheral nervous system are surrounded by a thin outer membrane, the **neurolemma,** which surrounds a series of flattened cells called **Schwann cells.** Most axons are also covered by a protective layer of lipids and proteins called the **myelin sheath.** As an axon develops, the sheath is actually secreted by the Schwann cells by concentric wrapping of the cell membrane. When the sheath is fully formed the outer most membrane of the Schwann cell forms the neurolemma. Schwann cells are located in series along an axon; their nuclei lie just underneath the neurolemma. Small gaps occur along the myelin sheath between adjacent Schwann cells. These gaps, termed **nodes of Ranvier,** are an important aspect of rapid nervous conduction. Axons of the central nervous system lack Schwann cells, but may still be covered in a myelin sheath. In these cases, the sheath is secreted by nerve fibers of specialized cells.

Nerve trunks and the white matter of the brain and spinal cord appear light in color because they are made up mostly of nerve processes covered by their white myelin sheaths. The nucleated nerve cell bodies are all located in ganglia or in the gray matter of the brain or spinal cord. The gray matter appears darker due to the lack of the sheath and the abundance of cell cytoplasm.

Examine a cross-section of a vertebrate nerve (spinal) cord. The vertebrate nerve cord is tubular and filled with **spinal fluid.** Note the tiny opening in the center of the cord. Locate the dark H-shaped or butterfly-shaped region in the center. This is the gray matter, containing the nerve cell bodies of motor and interneurons. The white matter contains both axons and dendrites that allow various parts of the nerve cord to communicate with each other and with the brain.

Several types of nerve cells are known. **Sensory neurons** carry impulses from a sensory ending, such as in the skin, to the central nervous system. **Motor neurons** carry impulses from the central nervous system to effector organs and the skeletal muscles. In the central nervous system, the links between sensory and motor neurons are called **interneurons.** These three neurons combine to form a **reflex arc** (see Chapter 10 for more information).

Nerve Impulse

A nervous impulse is actually a fleeting wave of changing electric polarity along the surface of a nerve fiber. Most cell membranes demonstrate a difference in electric charge between the inner and outer surface of the membrane. This differing electric charge creates a **membrane potential,** which in a nerve cell at rest is approximately 70 millivolts (the **resting potential**). At the specific position of a nerve impulse the membrane potential (+ outside; – inside) is reversed. This reversal is due to the sudden influx of sodium ions (Na^+) through ion channels in the membrane. The normal electric potential is quickly restored when the impulse passes as potassium ions (K^+) are pumped out. Immediately thereafter, during the **refractory period,** sodium ions are exchanged for potassium ions and the resting potential is restored.

Synapses

Junctions between axons and adjacent nerve cells form small gaps termed **synapses**. Synapses are regulatory structures that control the rates of impulse transmission from one neuron to another. There are two basic types of synapses: a **chemical synapse**, which relies on molecular exchange across the gap, and an **electric synapse**, which has a much narrower gap that ionic currents flow directly across. Synapses are the basis for integration between nervous systems; they function as gates that selectively determine which, and at what rate, nervous stimuli are passed on.

ORGAN SECTIONS

Tissues are are often highly integrated and interactive in advanced animals. Organized masses of tissue collectively form **organs**. The arrangement of tissue layers in an organ is related to function, and many variations exist. Examine several slides of organ sections and determine the type and number of tissue layers present. Note any associations between certain tissues (such as epithelial and connective) and the arrangements of each type in different organs. The vertebrate body has many complex organs with a variety of tissue arrangements; a few are described below and two examples of skin (amphibian and mammalian) are presented in the next exercise.

Cross-Section Through an Artery, Vein, and Nerve

Blood vessels, and arteries in particular, are muscular organs. Because an artery contains blood under higher pressure than that of a vein, arteries have a smaller diameter and thicker walls than the corresponding vein. Examine a cross-section slide of the blood vessels (Figure 7.18). The artery may collapse and appear oval and the thin muscular walls of the vein may be folded. Nevertheless, the artery wall will be much thicker than the venous wall. The innermost layer of both is a thin lining of **simple squamous epithelium**. The bulk of the wall is composed of a circular layer of **smooth muscle** which may bear some elastic fibers. The outer layer is made up of fibrous **connective tissue**. Near the vessels, locate one or several solid sections through nerve trunks. Dispersed across the slide will be some **adipose tissue** and **connective tissue** which functions in holding the vessels and nerves in place.

Cross-Section Through the Small Intestine

The digestive tract of higher animals consists of regionalized compartments which are adapted to perform different functions. All of these regions house an inner chamber, through which material passes, termed the **lumen**. Examine a cross-section slide of the small intestine, the primary digestive and absorptive area (Figures 7.19 and 7.20). Locate the open area, which is the lumen. The lumen is lined with an infolded mucous membrane, the **muscosa**, made of **columnar epithelium**. Some of these columnar cells are highly specialized and form secretory **goblet cells**. Note the vase-like shape of the goblet cells and compare their appearance with the more typical columnar cells. Underlying the mucosa is the **submucosa**, a layer of connective tissue containing collagenous fibers interspersed with small blood vessels.

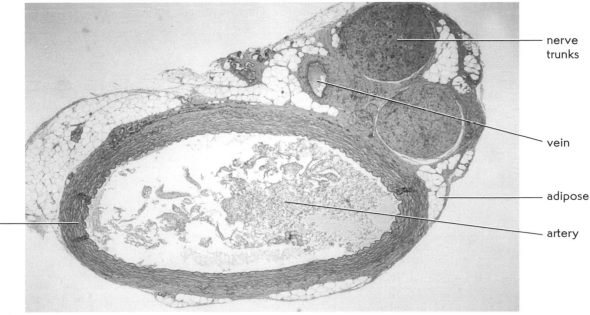

nerve trunks

vein

adipose

artery

smooth muscle

FIGURE 7.18 Cross-section through an artery, vein, and several nerve trunks. Note the presence of adipose tissue.

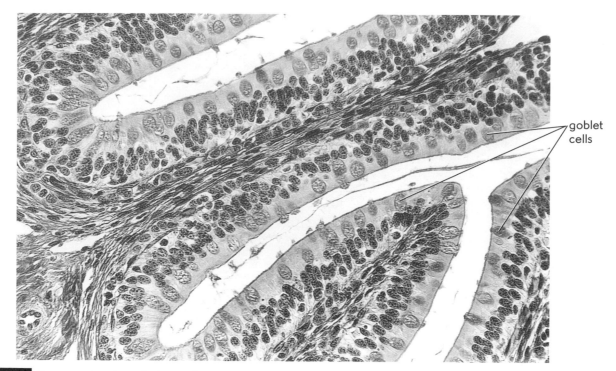

goblet
cells

FIGURE 7.19 Cross-section through the small intestine, showing an abundance of goblet cells.

Cross-Section Through the Trachea

The trachea, or windpipe, is a flexible tube that brings air to the bronchi on its way to the lungs. It is supported by open rings of **hyaline cartilage** (Figure 7.21).

The innermost layer of the trachea is a mucosal lining of **ciliated pseudostratified epithelium** which contains goblet cells and rests on a basement membrane of **connective tissue.** A submucosal layer contains mucous glands composed of **cuboid epithelium.**

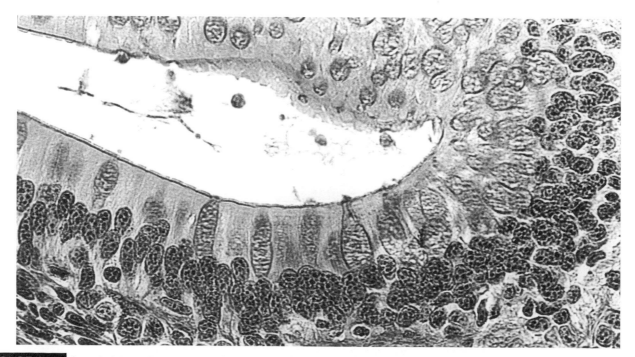

FIGURE 7.20 Detail of the columnar epithelium in the mucosal layer of the small intestine.

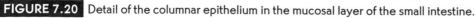

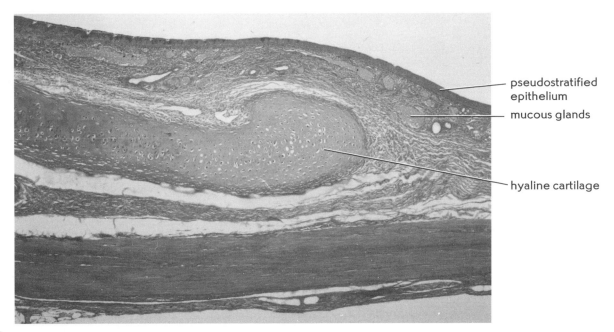

pseudostratified epithelium

mucous glands

hyaline cartilage

FIGURE 7.21 Cross-section through the trachea.

✓ Checklist of Suggested Demonstrations

TISSUE STRUCTURE AND FUNCTION

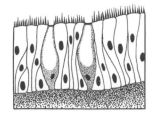

_____ 1. Simple squamous epithelium.

_____ 2. Simple cuboidal epithelium.

_____ 3. Simple columnar epithelium.

_____ 4. Stratified squamous epithelium.

_____ 5. Stratified columnar epithelium.

_____ 6. Pseudostratified columnar epithelium.

_____ 7. Areolar connective tissue.

_____ 8. Adipose connective tissue.

_____ 9. Hyaline cartilage connective tissue.

_____ 10. Compact bone cross-section.

_____ 11. Human blood smear.

_____ 12. Lymph gland longitudinal section.

_____ 13. Smooth muscle.

_____ 14. Skeletal muscle.

_____ 15. Cardiac muscle.

_____ 16. Nerve longitudinal section.

_____ 17. Spinal cord cross-section.

_____ 18. Artery and vein cross-section.

_____ 19. Small intestine cross-section.

_____ 20. Tracheal cross-section.

Tissue Notes and Drawings

Tissue Notes and Drawings

EXERCISE 8

Vertebrate External Anatomy, Skeleton, and Muscles

The anatomy of vertebrates is quite diverse, yet homologous features and structural precursors are readily observable in many vertebrate groups. Fish, primitive tetrapods (amphibians and reptiles), birds, and therian mammals show a range of increasing complexity and advancement. However, for most systems, the basic vertebrate architectural pattern is well represented by a common amphibian, the frog. For comparative study, a bony fish and a fetal pig commonly serve as representative vertebrates. This exercise, and the following one, focus on the frog as a representative vertebrate. Your instructor may include comparative demonstrations utilizing other vertebrates, such as a fetal pig.

One of the most common frogs in North America is the leopard (or grass) frog, *Rana pipiens*. Another frog, which is often studied because it attains a larger size, is the bullfrog, *Rana catesbeiana*. Either species can be used in this exercise, depending upon availability.

EXTERNAL ANATOMY OF THE FROG

Obtain a preserved specimen of either *Rana pipiens* or *Rana catesbeiana* and locate the following external features. The body of a frog consists of an anterior **head** and a posterior **trunk** region (Figure 8.1). The adult frog lacks a post-anal tail; their aquatic larvae, tadpoles,

retain such a tail, but it is absorbed, and the cells recycled (through the hydrolytic activity of lysosomes), during metamorphosis. The adult frog is adapted to live in environments which are transitional between land and water. Hence locomotory responsibilities are shifted from the tail to the long, powerful **hindlegs**. In fact, much of the adult frogs anatomy is geared toward elongation of the pelvic girdle and the hindlimbs, a modification to increase jumping leverage. The front legs are somewhat reduced used primarily to keep the head raised from the ground and to aid in external fertilization (they allow the male to secure the female as she releases her eggs). Note the absence of a distinct neck; the wide head is fused to the trunk. This is due to the retention of a primitive characteristic of fishes, for which an independent motion of the head and trunk would be disadvantageous during swimming. The head is dominated by the large **mouth** which extends nearly the entire width of the skull. This is a feeding adaptation; frogs are generalists which will prey on nearly any item which they can swallow. On top of the head locate the **external nares** and the eyes. These sensory organs are placed on top of the head to allow efficient operation while the frog is partly submerged.

The upper **eyelid** is closeable and formed from a simple fold of skin. The lower lid is a transparent **nictitating membrane** that can be drawn across the surface

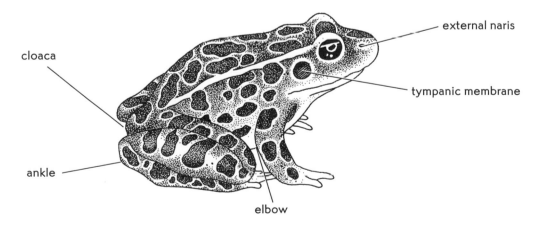

FIGURE 8.1 Lateral view of *Rana pipiens*, the leopard frog, with external structures indicated.

of the eyeball. The frog uses this membrane to moisturize and clean the surface of the eye as well as for protected vision while underwater. Posterior to each eye is a flat spherical area which receives soundwaves. This eardrum, or **tympanic membrane** is delicate and easily punctured. Although it is the same size in male and female leopard frogs, it is considerably larger in male bullfrogs. Carefully examine the top of the head between the eyes, and locate a small, light **brow spot,** which is about the diameter of a pin. This is a vestige of a light-sensitive eye that characterized primitive groups of fishes and amphibians.

The common opening of the digestive and urogenital tracts, the **cloacal aperture,** is located at the posterior end of the trunk just above the junction of the hindlegs to the trunk. This opening discharges undigested wastes (as feces), liquid excretory products (urine), as well as reproductive cells (gametes).

Although adapted for jumping, the frog is a fairly typical tetrapod. The foreleg consists of an upper arm, an elbow, a lower arm (or forearm), a short wrist, and a hand with five digits. It may appear that there are only four fingers. This is due to a reduction of the **prepollux,** an appendage equivalent to the human thumb. In frogs, the finger closest to the body is often stout and may be referred to as the "thumb." During the breeding season, the "thumb" in the male thickens and becomes heavily pigmented. The hindleg consists of the thigh, the knee, the shank, the ankle, and the elongated foot. Two elongated ankle bones (the astragalus and calcaneus) lie within the proximal part of the foot: the distal part bears five long toes. The first of five digits, the smallest and closest to the body, is called the **hallux.** The small spur at its base is the **prehallux.** In the more terrestrial toads, the prehallux is expanded into a spade-like structure modified for burrowing.

VERTEBRATE SKIN

All organisms possess a protective external covering of the body. In advanced animals, such as vertebrates, the covering may be thickened and specialized for protection, preventing desiccation and providing insulation. Even unicellular organisms have an external membrane which functions in protection, osmoregulation, and may house specialized locomotory projections (flagella and cilia). The external covering of vertebrates is a multicellular layer, termed the **skin,** which comprises the **integumentary system.** In addition to the general functions mentioned above, the vertebrate integument forms a primary defense against invading disease organisms and may also be specialized for water absorption, and/or **cutaneous respiration.**

The integument of vertebrates consists of two layers: an external **epidermis** and an underlying **dermis.** The epidermis is composed of several layers of epithelial cells, of which the outermost layers are usually formed by dead, or **keratinized,** cells. Keratin is a protein

molecule that is used to waterproof the skin. The placement of keratin in these dead cell layers is a primary feature which has allowed vertebrates to successfully colonize dry terrestrial environments. Densely packed keratin layers are found in reptiles, birds, and mammals; the skin of amphibians is moist and glandular, thus increasing the efficiency of cutaneous respiration, and is only slightly keratinized. The dermis is composed of fibrous connective tissue. Next we will compare the structure of amphibian and mammalian skin sections to demonstrate some anatomical extremes of vertebrate integument.

Amphibian Skin

Examine a stained slide of a cross-section of frog skin (Figure 8.2). The outer epidermis is composed of **stratified squamous epithelial** tissue. This layer appears dark in cross-section because it contains a large amount of nuclei relative to cytoplasmic material. The underlying **dermis** is lighter in color and contains only a few nuclei. The cells of the outermost layer of the epidermis are typically thin, flat squamous cells. These are oriented in parallel fashion with the surface; the deeper epithelial cells are more cuboidal. The cells in the basal layer are usually longer than they are wide, hence they are columnar in shape. The boundaries between cells are often indistinct, but the spacing between and the arrangements of the nuclei reveal the shapes of the individual cells. The epidermis, with its successive gradation from superficial squamous cells to deeper columnar cells is produced by the basal **germinative layer** which continually produces new cells that are pushed out toward the surface and become steadily flatter and harder (more **cornified** or keratinized) as new cells are formed in the germinative layer. After long aestivatory or hibernation periods, and sporadically during times of peak activity and favorable environmental conditions, a new layer of skin forms beneath the old one. The older layer is then molted, or sloughed off and often eaten by the animal. Because epithelial tissue is thin and lacks a blood supply, you will not find blood vessels or nerves penetrating the epidermis.

The dermis is characterized by the presence of two types of prominent glands: the **mucous glands** and the **poison glands.** Both of these glands are **exocrine glands** (i.e., they release their products through ducts) that secrete materials to the surface of the epidermis. Both types of glands are lined by a layer of cuboidal secretory cells formed as an epithelial layer which descends down into the dermis during tissue development. The poison gland can be identified by its larger size and the presence of granular material within the lumen of the gland. If an amphibian is disturbed or frightened, the poison glands discharge a thick toxic secretion that can irritate the mucous membranes of a predator. This secretion is particularly strong in toads, and many animals will regurgitate the toad if it is consumed.

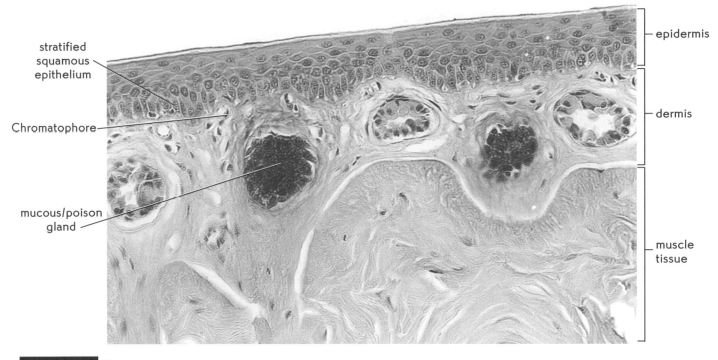

stratified squamous epithelium

Chromatophore

mucous/poison gland

epidermis

dermis

muscle tissue

FIGURE 8.2 Cross-section through frog skin.

Humans can be affected by the toxin if it enters the bloodstream. This occurs most often when the membranes of the eyes, nose, or mouth are contacted after handling an amphibian. Poison arrow frogs (*Dendrobates*) of South America produce extremely toxic alkaloids that are potentially fatal to humans. Some large toads of the neotropics (*Bufo marinus*; introduced to Australia, Bermuda, and southern Florida) produce enough toxin to kill a dog or cat if ingested.

The mucous glands produce a colorless, watery secretion that moistens the skin. The rate of secretion is regulated by the expansion or contraction of a **stoma cell** which is located at the surface opening of the duct. Move the slide around and locate a mucous gland with its accompanying stomal cell. These may be diffucult to find or lacking on your slide depending upon how the tissue was prepared.

Between the epidermis and the dermis, locate a dark layer of **chromatophores,** cells that contain pigment granules. Under the influence of light and hormones, these cells lighten or darken and can regulate the protective coloration of the animal. Many aquatic and semiaquatic animals, such as frogs, exhibit a permanent coloration pattern termed **countershading.** Countershading is designed to camouflage the animal against its background. To deep water animals, surface environments appear light and to animals at the water surface or on land looking down into the water, environments appear dark. Thus, the cryptically countershaded animal possesses a light ventral undersurface and the dark dorsal top surface. Most ranid frogs, including *Rana pipiens,* exhibit countershading (green on the dorsal and lateral surfaces and white or pale

coloration on the ventral surface). The patterns of most amphibians are stable, but the colors of some undergo drastic changes of intensity. The skin darkens when the pigment granules in the chromatophores spread out and cover other elements in the skin, and it lightens when they are concentrated. The movement of these pigment molecules within a chromatophore are produced by microtubule action. Changes in color can result from both external conditions and internal metabolic conditions; a low temperature produces darkening, whereas a high temperature, drying of the skin, or increased light intensity causes the skin to lighten.

The dermis contains a large number of small blood vessels that help regulate body temperature, transport nutrients and carbon dioxide to the skin and carry oxygen back to the heart. Nerves and muscle tissue are also present in the dermis but are probably not visible in this preparation.

Mammalian Skin

Mammalian skin is more complex than that of amphibian skin. Examine a slide of mammalian skin, preferably taken from an animal which possesses sweat glands (not all mammals do). The **epidermis,** while still composed of **stratified squamous epithelial** cells, is thicker than that of the frog and has several more layers of keratinized cells at the surface (Figure 8.3). The epidermis can be divided into several distinct layers: the outermost **stratum corneum;** an intermediate layer, the **stratum granulosum;** and a basal **stratum germinativum.** The stratum corneum consists of dead flattened cells filled with keratin. These cells are continuously

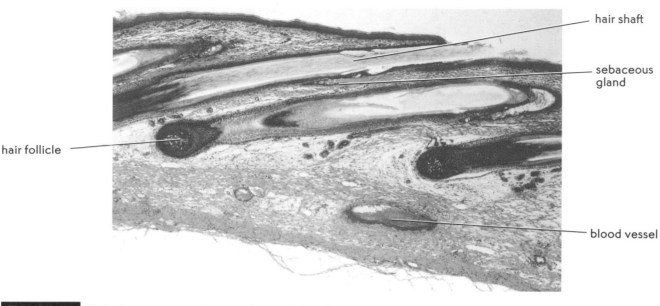

hair shaft

sebaceous gland

hair follicle

blood vessel

FIGURE 8.3 Vertical section through mammalian (pig) skin. Sweat glands are absent in pigs.

being sloughed off and replaced by the underlying layers. New skin cells form in the stratum germinativum and keratin is synthesized here, filling in the cytoplasmic space as the cells age and migrate toward the top of the epidermis. The epidermis and the dermis are separated by a thin layer of collagen fibers termed the **basal lamina.** The underlying **dermis** consists primarily of connective tissue which is heavily reinforced with interwoven fibers. The outermost layer of the dermis consists of loose connective tissue organized into the **papillary layer.** The papillary layer is convoluted and tufts of dermis push into the epidermal stratum germinativum. This arrangement allows blood vessels to closely associate with the epidermis, though they do not actually penetrate it. Below the papillary layer is an area of dense connective tissue composed of collagen and elastic fibers. This layer imparts both strength and the essential elasticity that the skin must have. Beneath the skin lies a layer of **subcutaneous fat** (adipose cells) which links the integument with the underlying musculoskeletal system. This fat layer is important in insulating the animal and is often utilized as an energy source.

Mammalian skin has a number of interesting specializations, many of which are epidermal derivatives which have descended down into the dermis during development. A unique mammalian feature is the presence of **hair,** a protective and insulatory device, on the surface of the skin. A single hair consists of a **shaft,** which penetrates through the dermis, and a basal **follicle,** which is embedded deeply into the dermis. Nerves, muscles, and blood vessels are all associated with each hair. **Sweat glands** are simple tubular exocrine glands which secrete watery fluids when the animal is actively thermoregulating. Because heat is absorbed from the surrounding molecules when evaporation (converting a

liquid to a gas) occurs, active sweat glands produce **evaporative cooling** of the skin surface. **Sebaceous glands** are always closely associated with a hair follicle. Sebaceous glands produce **sebum,** an oily secretion that helps lubricate the hair and skin surface. During cold weather sebum may retard evaporation and limit heat loss to the environment. **Pheromone** (scent) **glands** are also present in mammalian skin but these may not be visible on your slide, depending upon the species of animal used and the location from which the skin was removed. Note the abundance of small blood vessels within the dermis. The increased vascularity of mammalian skin is an important thermoregulatory feature that allows the animal to shunt blood to the skin for cooling.

VERTEBRATE SKELETON

The vertebrate **endoskeleton,** an internal structural support system, represents a great anatomical advancement over the hydrostatic and exoskeletal support systems of invertebrates. While the invertebrate plan is successful, it often limits the size of the animal and may restrict movement. The endoskelton, however, provides protection and structural support for the organs, but also allows for strong muscle attachments, flexibility of motion, and nearly unlimited growth potential. All animals which possess an endoskeleton have a basic skeletal organization composed of two major parts: the **somatic skeleton,** bones of the body wall and appendages, and a less conspicuous **visceral skeleton,** located chiefly on the wall of the pharynx. The visceral skeleton is important in fishes for gill support but it is reduced in higher vertebrates to the **hyoid apparatus,** a small bone and cartilage structure which supports the jaws, pharynx, larynx, and floor of the mouth. The somatic skeleton is subdivided into the **axial skeleton** and the **appendicular skeleton.** The axial skeleton contains bones of the skull, the vertebral column, the sternum, and and ribcage; the appendicular skeleton consists of the limb bones and their support systems, the pelvic and pectoral girdles.

Using the frog as a representative vertebrate, locate the axial and appendicular systems using Figure 8.4. Obtain a mounted frog skeleton and familiarize yourself with the components of the skeletal system. Some common bone terminology is presented to facilitate your understanding of the relationships between bone and the soft tissues. We will examine the frog skeleton in closer detail utilizing Figures 8.5, 8.6, and 8.7.

Bone terminology

Foramen: opening or passageway through a bone; usually designed to allow blood vessels or nerves to pass through the bone en route to a soft tissue or the bone itself (e.g., the **foramen magnum** allows the spinal cord to pass through the skull to the brain).

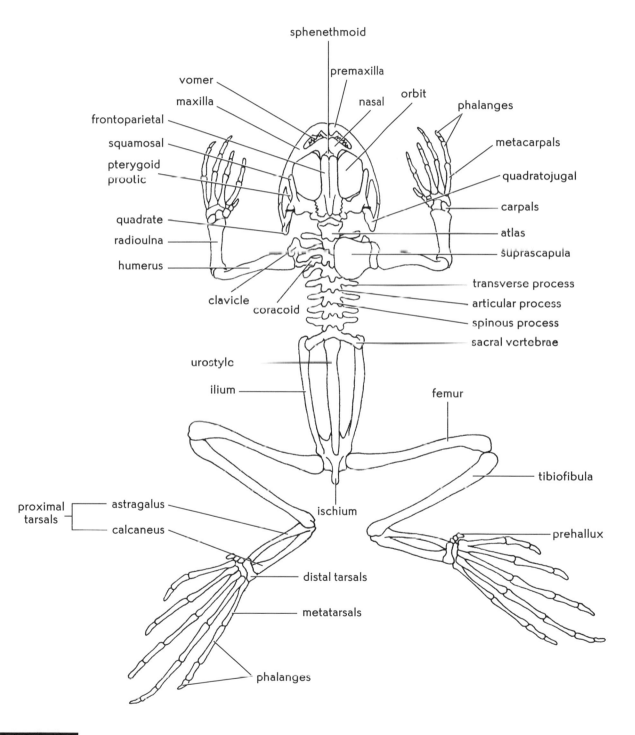

FIGURE 8.4 Dorsal view of the frog skeleton. The suprascapular cartilage has been removed on the left side of the drawing.

Fossa: an indentation or small cavity in a bone; increases the surface area for muscle attachment, thus strengthening the muscle (e.g., the **mandibular fossa** of the lower jaw allows the **depressor mandibulae** muscle to strengthen its attachment to the mandible).

Fenestra: a small opening, or window, through a bone; a fenestrated bone can have many pores or perforations; fenestrae are often passageways (e.g., the **fenestra cochleae** allows the release of vibrations to the inner ear in mammals).

Condyle: articulating surface of a bone which connects to an adjacent bone to form a joint (e.g., the **exoccipital condyle** of the skull articulates with the first cervical vertebra, the **atlas,** of the neck).

Axial Skeleton

Skull

The **skull** houses the protective braincase (**cranium**) as well as the components of the visceral skeleton (the hyoid apparatus, the jaw cartilages, and the **columellae** — ear bones). Although the frog skull is broad, much of the skull is dominated by large vacuities (termed **fossae**) which reduce the weight of the skull and allow soft tissues to penetrate the skull. Both the **nasal** and the **orbital fossae** are readily apparent in dorsal view. The skull also protects the **olfactory, optic,** and **auditory capsules;** these are protective areas which respectively house the organs of smell, sight, and hearing. As noted earlier, chondrichthyan fish (sharks, skates, and rays) possess a cartilaginous skull, but all higher vertebrates have an ossified skull. A trend toward reduction in the number and complexity of skull bones is apparent as you progress through the vertebrate classes. Some fish have up to 180 skull bones; amphibians and reptiles, from 50 to 95; and mammals, 35 or less. Human beings have 29. The density of the skull varies depending upon functional need. Many small amphibians have a thin, hardly ossified skull; many birds have an extremely thin skull which facilitates flight by reducing the weight of the animal; and crocodiles and alligators have an extremely dense skull which is used to secure and crush large prey items.

Using the mounted specimen, examine the dorsal surface of a frog skull (Figure 8.5). The skull is attached to the anterior end of the **vertebral column.** Locate the small cranium, which is formed by the **frontoparietal** bones, the transverse **nasal** bones (which surround the nasal capsules), the posterior **prootics** (which house the inner ears), and the **exoccipital** bones (each of which has posterior articular surface, the **occipital condyles**). These two condyles fit snugly in depressions of the first vertebra, the **atlas,** permitting movement of the head relative to the vertebral column. A large opening, the **foramen magnum,** lies between the two occipital condyles. This is the opening through which the spinal cord passes through to the brain. Between the prootics and the **maxilla** (upper jaw) are the **squamosal** and **pterygoid** bones. Together these form the lateral borders of the braincase. The squamosal is a T-shaped bone which forms the posterior and dorsal elements of the jaw joint. The pterygoid is also T-shaped but is ventrally located. The upper jaw consists of the two small anterior **premaxillary** bones and the larger **maxillary** bones that extend posteriorly to join with the pterygoid. Both the premaxillae and the maxillae bear small **homodont teeth.**

The ventral surface of the skull reveals the **maxillary teeth** on the margin of the upper jaw and two large crushing **vomerine teeth** on the roof of the mouth (Figure 8.5). The **sphenethmoid** bone, which connects with the frontoparietal bone dorsally and the **parasphenoid** bone ventrally, make up the lateral and ventral part of the braincase. The **mandible,** or lower jaw, lacks teeth and is composed of two bones in the frog, the anterior **dentary** and the posterior **angulare** bones. The suture between these two bones can become obscure in older bullfrogs and may not be visible. The jaw joint is padded with a rod-like **Meckel's cartilage** (part of the visceral skeleton) which separates the lower jaw from the upper jaw.

Vertebral Column

The arrangement and number of vertebrae, as well as the development of their processes, varies greatly among different species and among different regions of the vertebral column in the same individual. The vertebrae in fish are differentiated into **trunk vertebrae** and **caudal vertebrae.** In most of the other vertebrates, they are further modified into **cervical** (neck), **thoracic** (chest), **lumbar** (lower back), **sacral** (pelvic), and **caudal** (tail) vertebrae. In birds, as well as in human beings, the sacral vertebrae become fused into a single bone called the **synsacrum.** This fusion strengthens the pelvic girdle and increases the possible weight load. The number of vertebrae varies for different animals. Some species of python have the most vertebrae — up to 435. Human beings have 7 cervical, 12 thoracic, and 5 lumbar vertebrae, plus 5 that have fused in the pelvic region to form the synsacrum and **coccyx.** The frog has a single cervical vertebra, 3 thoracic vertebrae, 4 lumbar vertebrae, and a single sacral vertebra.

Note that the vertebrae of the frog are very similar to each other. The lateral projections are called **transverse processes;** these are more prominent on the thoracic and lumbar vertebrae of the frog and culminate in ribs in other vertebrates. The general vertebrate plan calls for a pair of ribs for each vertebra from head to tail, but the tendency has been to reduce the number from the lower to the higher forms. Ribs, which are designed to protect the vital organs of the body cavity, however, are not common to all vertebrates; many, including the frog, do not have them at all. Human beings have 12 pairs of ribs.

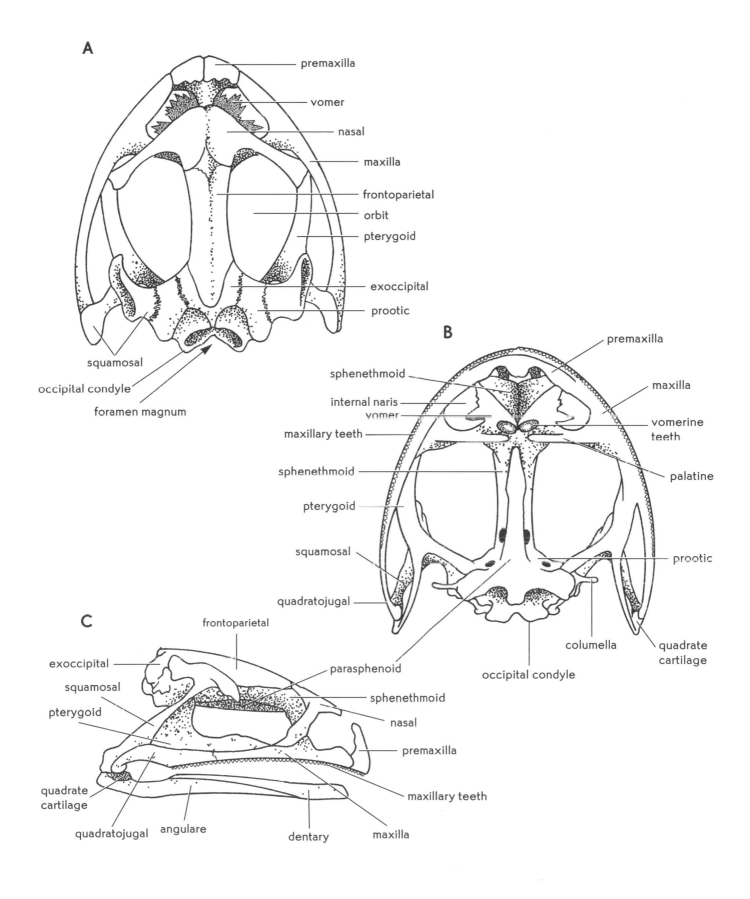

FIGURE 8.5 Frog skull. A) dorsal; B) ventral; and C) lateral views.

Appendicular Skeleton

Except for some highly adaptive burrowing species, most vertebrate animals have paired appendages, which are supported and linked to the spinal column by **pectoral** (shoulder) and **pelvic** (hip) **girdles** (Figures 8.6 and 8.7). Many vertebrates show structural modifications in the girdles, limbs, and digits, depending upon the specific needs of their lifestyle. The girdles and appendages of fishes are relatively similar as are those of most tetrapods. Both the cartilaginous and bony fishes have pectoral and pelvic fins, which are supported by the pectoral and pelvic girdles, respectively. Tetrapods usually possess a single proximal (closest to the midline of the body) limb bone which articulates with the respective girdles. In frogs and most other vertebrates, the proximal forelimb bone is the **humerus** and the proximal hindlimb bone is the **femur.** Distal (away from the midline of the body) to these are the **radioulna** (a fusion of the **radius** and the **ulna**) in the forelimb and the **tibiofibula** (a fusion of the **tibia** and **fibula**) in the hindlimb. The humerus is attached to the pectoral girdle in the deep **glenoid fossa** by ligaments. The pectoral girdle in the frog consists of several bones not found in higher vertebrates. The **scapula** is laterally placed in the frog and extends to a broad dorsal plate-like bone, the **suprascapula.** Ventrally, a small **clavicle** is present just above a larger **coracoid.** The coracoid is a typical amphibian feature first seen in fossil labyrinthodonts. Higher vertebrates have lost the coracoid. The junction of the two clavicles and the two coracoids form a small central **sternum,** which continues anteriorly as the **omosternum** and **episternum** and posteriorly as the **mesosternum** and **xiphisternum.**

The pelvic girdle is composed of three fused bones: the **ilium,** the **ischium,** and the **pubis.** The arrangements of these three bones differs between some groups of vertebrates and is a primary distinguishing characteristic between two large groups of dinosaurs, the ornithischians (bird-hipped dinosaurs) and the saurischians (lizard-hipped dinosaurs). In frogs, the femur attaches to the pelvic girdle in a socket, the **acetabulum.** The feet and hands are built according to a common pattern, with a number of **carpal** (wrist) or **tarsal** (ankle) bones attached to a group of hand (**metacarpal**) or foot (**metatarsal**) bones, followed by the **phalanges,** bones of the digits (fingers and toes).

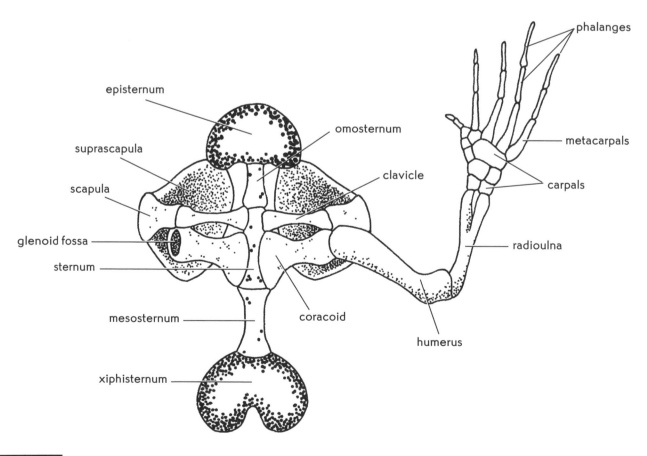

FIGURE 8.6 Pectoral girdle and anterior limb bones of the frog.

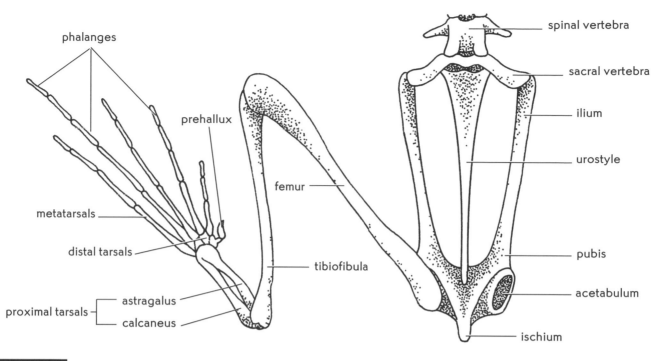

Labels in figure:
phalanges
prehallux
metatarsals
distal tarsals
proximal tarsals — astragalus / calcaneus
femur
tibiofibula
spinal vertebra
sacral vertebra
ilium
urostyle
pubis
acetabulum
ischium

FIGURE 8.7 Pelvic girdle and posterior limb bones of the frog.

VERTEBRATE MUSCULATURE

Before examining the muscular system of the frog, we will cover some basic terminology associated with different joints and muscle actions.

Joint Terminology

Ball-and-socket: allows movement in any direction, including rotation [e.g., the shoulder joint (**humerus** and **glenoid fossa** of the **coracoid**), and the hip joint (**femur** and **acetabulum**)].

Condyloid: allows movement in any direction except rotation (e.g., joint between **metacarpals** and **phalanges**).

Hinge: allows bones to bend in only one direction (e.g., the knee joint formed by the distal end of the **femur** and the proximal end of the **tibiofibula**).

Plane: allows sliding movements between flat surfaces on bones (e.g., between the **vertebrae**).

Radial: permits rotation of one bone on another [e.g., between the proximal ends (ends nearest the humerus) of the **radius** and **ulna**].

Muscle Terminology

Muscles accomplish several different tasks for the body. They support the body by maintaining the proper relationship between articulating bones; they are responsible for all bodily movements, including movement of the body itself and the movement of internal fluids, gases, and materials. Skeletal muscles are grouped into fibrous bundles which attach to a bone or other muscles by cord-like, connective tissue termed a **tendon** or by a broad, connective tissue sheath called an **aponeurosis.** One end of a muscle is usually fixed in position; this nonmovable end is called the **origin**. The opposing end of the muscle is termed the **insertion**. Upon contraction of the muscle, the insertion moves pulling a bone or other structure toward the origin. Many skeletal muscle bundles are fusiform or spindle-shaped with a broad central region, the **belly**, located between the two fibrous tendons.

The only action a muscle can undergo is contraction, or shortening of the fibers. This causes two structures (usually bones) to be brought together. To return to the resting position, an **opposing** or **antagonistic muscle** must then contract. Thus, skeletal muscles are commonly arranged in antagonistic pairs, where one muscle causes movement in one direction and an opposing muscle causes movement in the opposite direction. The following list describes a few common types of opposing muscles:

Flexors cause bending, by decreasing the angle between two structures (e.g., the biceps muscle bends the forearm toward the upper arm).

Extensors straighten or extend a part of the body (e.g., the triceps muscle extends the forearm away from the upper arm).

Adductors move a part towards the axis or midline of the body (e.g., the latissimus dorsi muscle draws the arm back against the body).

Abductors move a part away from the axis or midline of the body (e.g., the deltoid muscle draws the arm away from the body).

Depressors lower a part (e.g., the depressor mandibulae muscle moves the jaw down to open the mouth).

Levators raise or elevate a part (e.g., the masseter muscle raises the jaw to close the mouth).

Removing the Frog Skin

Because cutaneous respiration is an important mechanism for gas exchange in frogs, the skin is highly vascular and not rigidly attached to the underlying muscles. Most vertebrates possess a thick layer of loose connective tissue (**fascia**) which surrounds muscle bundles and secures the skin. Frog skin is only rigidly attached in several important areas (such as at the joints, on the head, around the cloaca and near the sensory organs). To expose the muscle groups, you will need to remove the skin in its entirety. First, locate a slit cut into the ventral (abdominal) surface of the frog. This is an opening created when the circulatory system was injected with latex to facilitate study of the blood vessels. Grasp the frog and insert a pair of scissors into the slit. Pull up on the skin in this region and carefully cut down the center of the body, making sure you do not cut any muscles which lie below the skin. Continue cutting down the midline of each leg until you reach the feet. Now return to the original entry point and cut the skin down each arm until you reach the hands. To quickly remove the entire skin, you may also want to cut in a circular pattern around the waist until you have circumvented the animal. You should now be able to peel the skin off inside-out, similar to the way you remove surgical gloves. Do not be afraid to pull hard, the skin is fairly tough and is resistant to breaking. When you have separated the entire skin from the trunk region, then pull it off the head. The skin is tightly bound near the eyes and tympanic membrane; keep pulling, it won't rip. After you have removed the skin, note the prominence of the small blood vessels on the undersurface of the skin. This is where oxygen diffuses into the bloodstream from the skin.

To expose the muscle groups underneath the skin, you will need to separate the adjacent bundles from each other. Individual bundles can be identified by the direction in which the fibers run. The fiber direction may be obscured by the fascia; you can remove this by prodding the edges of the bundle with a blunt probe and peeling the fascia from the muscle surface. Now take your blunt probe and run the probe along the sides of the muscle bundle. If the probe runs cleanly and easily from tendon to tendon then you have successfully isolated a muscle bundle. If the fibers become exposed and the muscle appears shredded, then you have inserted the probe into a bundle and torn the fibers. Continue isolating the obvious bundles and use Figures 8.8 and 8.9 to guide you. The fascia of the lower back (dorsal surface) is thicker than that of other areas; you will need to carefully dissect away the fascia

in this region to expose the underlying muscles. Muscles of the following body regions will be examined: the head, shoulder, forelimb, trunk, thigh, and shank of the hindlimb. We will mainly focus on superficial (surface) musculature.

Muscles of the Head

Muscles on the dorsal surface of the head include the **temporalis**, the **masseter**, the **pterygoideus**, the **cucullaris**, and the **depressor mandibulae**. Most of these muscles are associated with movements of the head. The depressor mandibulae inserts on the lower jaw and during contraction it opens the mouth (depresses the mandible). The cucularis is a small muscle located on the posterior edge of the tympanic membrane just underneath the depressor mandibulae. The cucularis is reduced in higher vertebrates and is retained in amphibians as a gill support remnant derived from their fish ancestors.

The ventral surface is covered by two transverse muscles, the **mylohyoid** and the **subhyoid**. Both of these muscles raise the floor of the mouth during swallowing and breathing (due to the positive pressure inhalation system of lower vertebrates).

Muscles of the Shoulder

The muscles of the shoulder are usually associated with movement of the arm relative to the trunk. The **dorsalis scapulae**, which lies beneath the depressor mandibulae, and the **latissimus dorsi** extend from the scapula, across the shoulder to the proximal end of the humerus (Figure 8.8). All these dorsally located muscles move the forelimb toward the midline of the body. Ventrally, locate the **deltoid**, the **pectoralis**, and the smaller **coracobrachialis** muscles which extend from the pectoral girdle to the humerus (Figure 8.9). These three muscles move the forelimb away from the body. The **coracoradialis**, another ventral shoulder muscle, extends from the girdle to the radioulna and is the major flexor of the forearm.

Muscles of the Forelimb

The **anconeus**, situated on the dorsal surface of the humerus, is the major extensor of the forearm. A number of small muscles extend the hand and digits. These include the: **extensor carpi radialis**, the **abductor indicis longus**, the **extensor digitorum communis longus**, and the **extensor carpi ulnaris**. The flexor muscles that bend the forearm toward the body are the **flexor carpi radialis**, the **flexor carpi ulnaris**, and the **palmaris longus**.

Muscles of the Trunk

Two relatively small muscles attach to the skin in the trunk region. Dorsally, the **cutaneous abdominis** extends

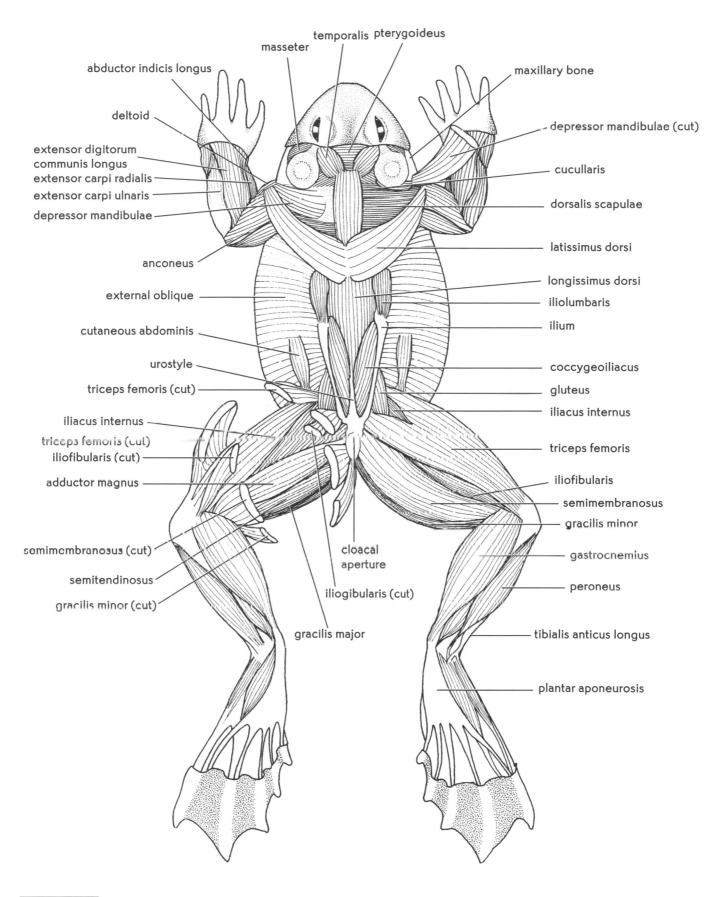

FIGURE 8.8 Dorsal view of frog musculature. Superficial muscles are on the right side of the drawing, deeper muscles on the left.

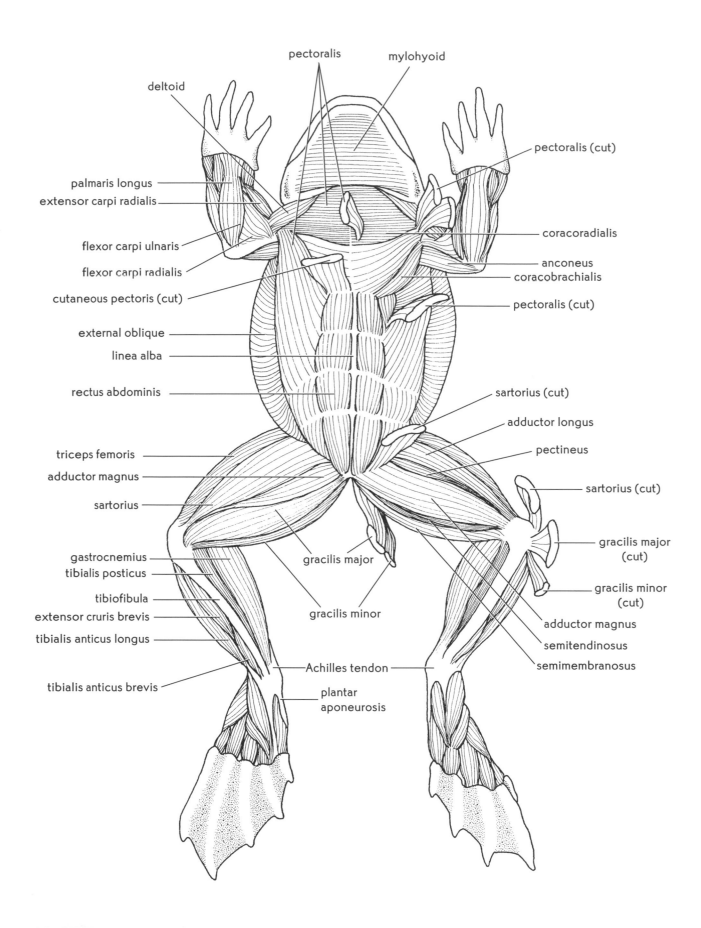

deltoid

pectoralis

mylohyoid

pectoralis (cut)

palmaris longus
extensor carpi radialis

coracoradialis

flexor carpi ulnaris

anconeus
coracobrachialis

flexor carpi radialis

cutaneous pectoris (cut)

pectoralis (cut)

external oblique

linea alba

rectus abdominis

sartorius (cut)

adductor longus

pectineus

triceps femoris

adductor magnus

sartorius

sartorius (cut)

gastrocnemius
tibialis posticus

gracilis major
(cut)

tibiofibula

gracilis minor
(cut)

extensor cruris brevis

tibialis anticus longus

gracilis major

adductor magnus

gracilis minor

semitendinosus

tibialis anticus brevis

Achilles tendon

semimembranosus

plantar
aponeurosis

FIGURE 8.9 Ventral view of frog musculature. Superficial muscles are on the left side, deeper muscles are on the right.

Introductory Zoology Laboratory Guide

from the legs and attaches to the large, lateral **external oblique** muscles. Ventrally, the anterior abdominal region is secured to the skin by the flap-like **cutaneous pectoris**. The contraction of these muscles may compress several lymph sacs and aid in lymphatic fluid circulation. The **longissimus dorsi** and **iliolumbaris**, located on the dorsal side of the trunk help strengthen and move the spinal column. Located within the framework of the pelvic girdle, is the **coccygeoiliacus** muscle. This muscle is attached at one end to the urostyle and at the other to the ilium bones. This muscle extends the urostyle and pelvic girdle to increase jumping leverage.

The ventral body wall of the frog is largely supported by the centrally located **rectus abdominis** muscles and the more lateral **external** and **internal obliques**. These three muscles serve to support and secure the abdominal region. A thin white area of connective tissue, the **linea alba,** connects the rectus abdominis muscles. If the frog has been injected with latex, this line may appear blue, due to the presence of the ventral abdominal vein just below the surface.

Muscles of the Thigh

The largest and strongest muscle of the frog body is the **triceps femoris,** so named because it possesses three "heads" or origins which attach to the ilium and it runs along the femur. The insertion is on the proximal end of the tibiofibula. This muscle extends the lower shank of the leg during jumping or swimming. The triceps femoris extends across the leg from dorsal to ventral surface, so parts of it can be seen in both views. On the ventral surface, below the triceps femoris, locate the **sartorius,** the **adductor magnus**, the **gracilis major,** and the **gracilis minor.** The sartorius is a thin "strap" muscle which crosses the medial edge of the triceps femoris and inserts on the tibiofibula. The adductor magnus originates on the pelvic girdle and inserts on the femur, thus adducting the thigh. The gracilis major and minor muscles both originate on the pubis and insert on the tibiofibula. Contraction of these two muscles causes extension of the thigh and flexion of the shank. Look

carefully for the thin gracilis minor. If it is not readily visible, check the skin to see if it was accidently removed.

Flip the frog over to its dorsal surface and locate the large **semimembranosus.** This is the dorsal complement to the gracilis major. Just above it, between the semimembranosus and the triceps femoris, locate the thin **iliofibularis.** This muscle extends and adducts the thigh and flexes the shank.

To see the **pectineus,** a deep muscle which runs along the femur, pull the sartorius up, and the adductor magnus down with a blunt probe. Look internally to observe a small muscle with fibers running parallel to the axis of the femur. Another important thigh muscle can be revealed by cutting the gracilis major and minor, along with the semimembranosus, across their bellies. Pull the ends back and locate a bifurcated cord-like muscle that runs from the pelvic girdle to the tibiofibula. This is the **semitendinosus,** a muscle which extends and adducts the thigh while flexing the knee.

Muscles of the Shank

The large "calf" muscle of vertebrates is the dorsal **gastrocnemius,** an important muscle in flexion of the shank and foot. The gastrocnemius terminates in the large **Achilles tendon** of the foot and originates as two thick tendons off the distal end of the femur. On the foot, the Achilles tendon broadens into a sheet of connective tissue called the **plantar aponeurosis.** Adjacent to the gastrocnemius, in dorsal view, is the **peroneus.** The ventral surface of the shank is dominated by a number of thin muscles running parallel to the tibiofibula along the lateral side. The **tibialis anticus longus,** a foot extensor, is the largest of the bunch. It runs the entire length of the shank. The shorter **tibialis anticus brevis** arises on the distal half of the tibiofibula and inserts on the tarsal bones. The **extensor cruris brevis** extends the shank and inserts on the proximal region of the tibiofibula. The thin **tibialis posticus** lies on the medial side of the tibiofibula, adjacent to the gastrocnemius.

EXERCISE 9

Vertebrate Digestive, Respiratory, Circulatory, and Urogenital Systems

This exercise covers basic vertebrate "soft" or internal anatomy, focusing on the primary organ systems: digestive, respiratory, circulatory, and the combined urinary and reproductive (urogenital) systems. We will again choose the frog as a representative vertebrate; your instructor may also dissect alternative animals (such as sharks, fish, mudpuppies, or fetal pigs) to provide comparative examples. Obtain the frog that you skinned in the last exercise; we will use the same specimen for further dissection.

DIGESTIVE SYSTEM

Buccal Cavity

Grasp your frog and pry open the mouth with a blunt probe. Work the probe to the corner of one jaw and insert the pointed end of your dissection scissors (Figure 9.1). Cut the hinge of the jaw until you have completely separated the tissue to approximately 1 cm behind the jaw joint. Repeat this procedure on the other jaw. The lower jaw should now be completely disengaged from the skull, and the mouth should open easily. Examine the **oral (buccal) cavity** and locate the broad **tongue** which lies on the floor of the cavity (Figure 9.2). Despite popular belief the tongue is not long and coiled, but broad and flat with a forked end. The tongue is attached anteriorly and is covered with a sticky secretion produced in the roof of the cavity. When prey is sighted, the tongue quickly flips out, extending the forks and grasping the prey. The prey item is then pressed against the tongue as the tongue is retracted into the oral cavity. Posterior to the tongue, the cavity constricts into the **pharynx,** which opens to the **esophagus.** The pharyngeal region is characterized by folded, vascular tissue which functions as an accessory respiratory surface. Between the tongue and the pharynx, locate a small longitudinal slit, the **glottis,** on the floor of the cavity. You may need to probe this area to reveal the glottis. The glottis is the opening into the pulmonary respiratory system; it opens for breathing but closes when food or water is in the cavity. In male specimens (you will determine the sex of your frog later

in this exercise) two small openings lie on the cavity floor on the sides of the pharyngeal region. These are the openings to the **vocal sacs,** structures which allow male frogs to call to the females during the reproductive season.

On the roof of the mouth, locate the **maxillary teeth** on the margin of the upper jaw. These are short chisel-shaped homodont teeth designed only for securing the prey, rather than mastication. Near the anterior edge, just posterior to maxillary teeth, locate the **internal nares,** and the **vomerine teeth.** The internal nares are openings which communicate with the external nares for respiration and the conduction of chemosensory information, via the olfactory nerve, to the olfactory lobe of the brain. The vomerine teeth are actually bony extensions of the vomer bone; these are designed for holding and crushing prey. At the back of the mouth, observe two paired openings to the **eustachian tubes,** each of which connects to the chamber of the middle ear beneath the tympanic membrane. Using a

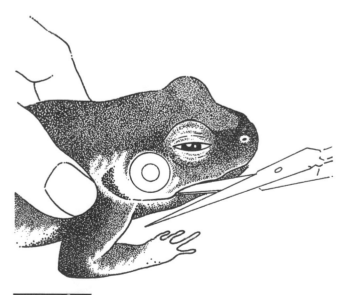

FIGURE 9.1 Dissection procedure of the frog mouth.

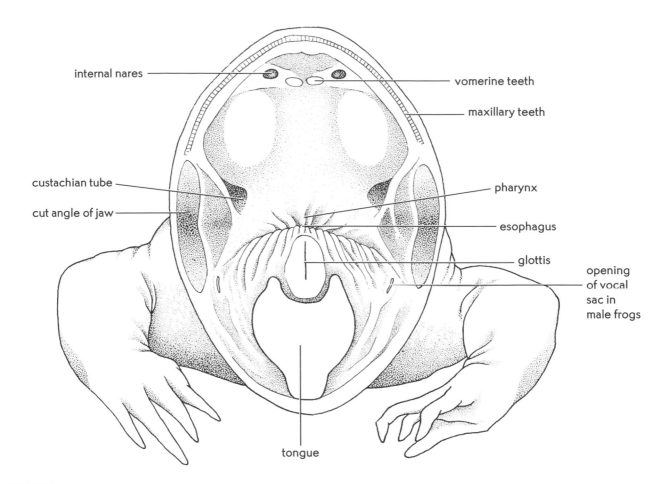

internal nares

vomerine teeth

maxillary teeth

custachian tube

pharynx

cut angle of jaw

esophagus

glottis

opening of vocal sac in male frogs

tongue

FIGURE 9.2 Internal view of the buccal (oral) cavity of a male frog. The angles of the jaw have been cut to allow the mouth to be opened widely.

needle probe you can easily determine the relationship between the eustachian tube openings and the external tympanic membrane.

Body Cavity

To expose the viscera, or internal organs, insert one end of your scissors into the small ventral opening, located in the chest region, which was created during preparation of the specimen. Cut longitudinally (just off center) through the abdominal wall, until you reach the pelvis. Now reinsert the scissors and cut anteriorly, being sure to cut completely through the sternum to the base of the head. Be sure and cut only the outer muscular layer of the body wall; a deep cut through the abdomen will damage the internal organs. Now make a transverse incision across the body wall just below the arms and just above the pelvis. You should be able to pull back the body wall of the entire abdomen, exposing the body cavity. The body cavity, or **coelom,** houses most of the important viscera. In mammals, the body cavity is divided by the **diaphragm,** a muscular structure which functions in ventilation. The anterior cavity, or **pleural cavity,** houses the lungs and surrounds the

heart, which sits in its own **pericardial cavity.** Beneath the diaphragm, the **abdominal cavity** contains the majority of the internal organs. Because amphibians lack a diaphragm, the coelom of the frog is divided into two slightly different cavities: the **pleuroperitoneal cavity,** containing the lungs and most of the other viscera, and the smaller pericardial cavity, containing the heart. In all vertebrates, the epithelial lining surrounding the heart is called the **pericardium.**

In eucoelomates, including frogs, the coelom is lined by two complete membranous epithelial layers, the **parietal peritoneum** and the **visceral peritoneum.** The parietal epithelium lines the body wall; the visceral peritoneum covers the surface of the organs. Both of these layers serve to protect the underlying structures as well as to isolate the structure and prevent contact with the **coelomic fluid.** The coelomic fluid, in turn, insulates the internal organs from drastic temperature shifts and cushions the organs from impact. Membranous sheets of epithelial tissue, called **mesenteries,** extend from the dorsal body wall to the organs as well as between many of the organs. The mesenteries serve as pathways for blood vessels and nerves as well as to inhibit excess movement of the organs.

Visceral Organs

Look into the body cavity of your frog. If you have a male specimen or a nonbreeding female, the body cavity should be full of elongate, yellow projections with an oily texture. These projections are the **fat bodies,** reservoirs for stored energy. The fat bodies extend from a dorsal location near the kidneys that will become apparent as you remove the internal organs. Young frogs, individuals under stress, and breeding females will not show extensive development of the fat bodies. Remove the majority of the fat bodies, noting the oily texture typical of lipid storage. Next, in the middle of the pleuroperitoneal cavity, locate a large, lobe-like organ which may appear green or brown speckled with bright blue flecks of color. This is the **liver,** the largest and often most prominent organ in the vertebrate digestive tract. In frogs, the liver is a trilobed organ which occupies much of the anterior portion of the coelom (Figure 9.3). The liver is a primary digestive organ; it produces **bile,** an alkaline digestive secretion which emulsifies fats. It also helps to filter venous blood for bacteria and undigested nutrients. The bile is stored in a sac-like organ, the **gallbladder,** located between the median and right lobes of the liver.

Beneath the left lobe of the liver, identify the muscular **stomach,** which lies dorsal to the liver and to your right as you view the body cavity. The stomach is

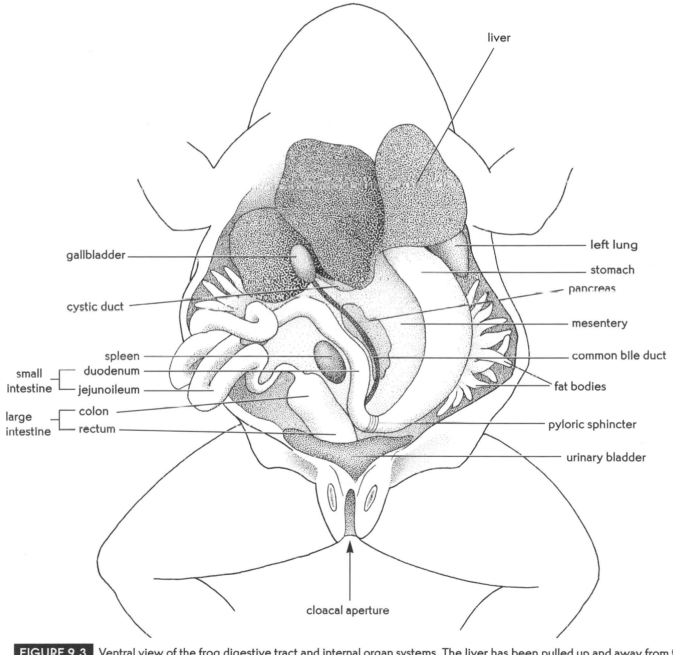

FIGURE 9.3 Ventral view of the frog digestive tract and internal organ systems. The liver has been pulled up and away from the abdominal cavity.

connected to the pharynx by the short **esophagus,** which lies hidden beneath the heart and lungs. The stomach is primarily a food storage organ that is responsible for some chemical and mechanical digestion of food. The passage of food into the stomach is regulated by the **cardiac sphincter,** a muscular band located at the anterior end of the stomach. As food is prepared for digestion and absorption it is broken down into a semisolid mixture called **chyme.** The passage of chyme into the intestines is regulated by the **pyloric sphincter,** a muscular area at the posterior end of the stomach. Using a scalpel, slit the stomach open longitudinally and locate the muscular grinding ridges (**rugae**) on the inner walls. If the stomach is contracted the rugae will be readily visible. If a large meal has recently been ingested, the stomach wall may be distended and the rugae indistinguishable. Probe the contents of the stomach and look for any undigested recognizable parts. Often the exoskeletal elements of arthropods (such as insects and crustaceans) pass through the stomach undigested. Fish are common prey items of aquatic frogs and you may be able to determine the meal content of your frog if digestion is not complete.

The stomach empties into the slender, highly coiled **small intestine,** the primary site of digestion and absorption. Note that the small intestine makes an immediate U-turn and ascends toward the anterior end. This is the **duodenum,** the first, or ascending, segment of the small intestine. Joining the duodenum immediately after the pyloric sphincter, is the **common bile duct.** This duct drains the gallbladder and the liver of their products and also services the **pancreas,** hence the term "common." In addition to producing an alkaline digestive secretion, the pancreas has microscopic clumps of endocrine tissue called the **islets of Langerhans;** these areas produce the hormones **insulin** and **glucagon,** both of which regulate blood sugar (glucose) levels. The pancreas is often difficult to distinguish in *Rana pipiens;* it is better developed in the bullfrog, *Rana catesbeiana.* Regardless, in frogs, the pancreas is always located in the mesentery between the **lesser (inner) curvature** of the stomach and the duodenum. In *Rana pipiens* the pancreas may be limited to a tissue thickening of the mesentery; in *Rana catesbeiana* the pancreas may appear as a lobe-like organ in the mesentery.

The remaining portion of the small intestine is the highly coiled **jejunoileum.** Notice the abundance of small blood vessels leading to the jejunoileum; this vascularity is an indication of the amount of nutrient absorption that occurs here. The jejunoileum terminates at the **large intestine,** a short, broad tube. The large intestine is differentiated into the anterior **colon,** the site of water reabsorption, and the posterior **rectum,** a waste storage area. Undigested waste products, or **feces,** are expelled through the **cloaca,** a common opening which services the digestive, urinary, and reproductive tracts.

Pull the stomach up and to your left. Embedded in the mesentery under the stomach, locate a small, round, brown organ, the **spleen.** The spleen is primarily a circulatory organ which serves as an important site for the production of red and white blood cells and the storage and destruction of aging red blood cells. It is also responsible for the production of antibodies (defense proteins produced in response to an antigen).

To properly complete the dissection you can now begin to remove selected organs and pick away the mesentery. First, remove as much of the fat bodies as possible, taking care to note their origin and perhaps keeping a small piece attached. Now cut away the bulk of the liver, leaving only a small section near the gallbladder for reference. Pull out the loops of the small intestine and carefully tear the mesentery supporting the coils. At all times during the dissection, be careful not to cut any blood vessels. They are injected with latex and will take some stress but not much. Use your blunt probe to tear the mesentery; as you do so, the intestine and other organs should become more pliable. Continue removing the mesenteries and isolating the organs until the body cavity appears less crowded and individual organs are easily identifiable.

In frogs the urinary and reproductive tracts are often closely associated with each other, forming the urogential system. The organs of the urogential system are located dorsally relative to the digestive tract. In the male frog, locate a pair of small, white, oval bodies, that are embedded into the mesentery on either side of the spinal cord; these are the **testes.** The female **ovaries** are located in the same relative position as the testes are in the male. The ovaries, however, vary greatly in size, depending on the animal's reproductive state. Nonbreeding females or immature females will possess a whitish gland near the junction of the fat bodies. The ovaries of breeding females are greatly expanded and take up all of the available space in the pleuroperitoneal cavity. The ovaries connect to a convoluted tube, the **oviduct,** through which eggs pass during ovulation and fertilization. Males may have a small vestigial oviduct. Dorsal to (underneath) each gonad (testis or ovary) is an elongated **kidney.** The kidneys are actually situated behind the parietal peritoneum, a condition referred to as **retroperitoneal.** The kidneys filter venous blood for toxins and impurities and concentrate this material as nitrogenous waste. Some vertebrates excrete this waste as a liquid (ammonia or urea) while a few others (birds and many reptiles) secrete it as a solid (uric acid). The kidneys are drained by a small tube, the **ureter,** into a storage organ, the **urinary bladder,** which sits at the base of the pelvis. The bladder may be collapsed in your specimen.

RESPIRATORY SYSTEM

Amphibians, and frogs in particular, respire by several different means. Shortly after hatching, the tadpole develops **gills,** which function in the absorption of dissolved oxygen from the water. This **branchial respiration** lasts until the tadpole begins to metamorphose into an adult frog. In both the adult and the tadpole stages oxygen is absorbed and carbon dioxide is eliminated through the skin (**cutaneous respiration**). Cutaneous respiration occurs both in water and in air and is a primary source of gas exchange for the animal. In fact, some salamanders lack lungs completely and must respire only in this fashion. During the winter, when the frog lies buried in the mud, the skin becomes the sole respiratory organ. Active adult frogs also absorb oxygen and give off carbon dioxide through the lining of the mouth (**buccopharyngeal respiration**) and through the lungs (**pulmonary respiration**). Although the lungs are considered the primary respiratory organ of most higher vertebrates, even in adult amphibians three or four times as much oxygen can be absorbed through the skin as through the lungs. In any case the respiratory tissues of vertebrates, particularly the skin of amphibians, must remain moist for gas exchange to occur.

During pulmonary respiration, air enters the body through the external nares and then passess into the oral cavity via the internal nares. The external nares are then closed and the floor of the mouth is raised, forcing air through the glottis into a short tube, the **larynx**. The larynx is reinforced by cartilage and, in males, contains two elastic bands, the **vocal cords,** that vibrate when air is forced from the lungs, resulting in a croaking sound. The larynx is connected to each lung by the **bronchus.** Note that the lung is not an empty sac, but consists of many internal pockets, called **alveoli,** which greatly increase the absorptive surface area of the lung. The alveoli are richly supplied with capillaries and this is the site of gas exchange in pulmonary respiration. Insert a small pipette into the glottis of your frog and pump air into the lungs. The lungs may be somewhat rigid due to the preservative and may not expand, but if they do note their great expansive capability.

CIRCULATORY SYSTEM

The circulation of fluids throughout the body is a primary functional need of all animals. In vertebrates, the circulatory system not only transports fluids and nutrients to cells and tissues but also provides a few other key functions: the transportation of respiratory gases; the transfer of stored materials; the elimination of waste, water, and excess minerals; the delivery of hormones and endocrine gland secretions; and the movement of antibodies and leukocytes. Several different circulatory systems exist in animals; the more primitive animals move fluids and materials through simple interstitial diffusion. Because diffusion through large areas is inefficient, advanced animals require a complex circulatory system to move materials through the bulk of their tissues. Most vertebrates possess a **closed circulatory system** with a **pumping organ** (the heart) designed to generate pressure to move the fluids. A closed system retains the fluids in enclosed **vessels** which direct fluids along definite routes. In vertebrates, the

fluid medium of the circulatory system is called **plasma.** Plasma consists mainly of water (90–92%) with some proteins (albumin; about 5–6%), and a few trace elements and ions. A number of cells, or corpuscles, flow in the plasma. Collectively, the plasma and its suspended corpuscles form vertebrate **blood.**

Frog Heart

Amphibians possess a **three-chambered heart** that houses two receiving chambers, the **left** and **right atria,** and a single pumping chamber, the **ventricle.** The three-chambered heart is a structural advancement over that of the two-chambered (a single atrium and ventricle) heart of fish. In a three-chambered heart the venous and arterial blood are separated in the atria, but some mixing of oxygenated and nonoxygenated blood occurs in the common ventricle. The frog heart is located in the center of the animal, just above the liver and between both lungs. The frog heart sits in its own cavity, the **pericardial cavity,** which is bound by the **pericardial membrane.** To reveal the heart, carefully cut away this membrane. Locate the thin-walled left and right **auricles** (the tissue surrounding the atrial chambers) and the thick-walled **ventricle** (Figure 9.4). Because the walls of the auricles are so thin, relative to the walls of the ventricle, the blood within them imparts a dark coloration to the auricles. On the dorsal surface of the ventricle locate a stout, cylindrical vessel, the **sinus venosus.** This vessel receives blood from the **posterior vena cava** as well as from the left and right **anterior vena cava.** Identify a pair of smaller **pulmonary veins,** which carry oxygenated blood back from the lungs and enter the left atrium. In ventral view, the two branches of the **truncus arteriosus** arise from a single **conus arteriosus.** The truncus arteriosus give rise to three aortic arches which distribute oxygenated blood to the **systemic circuit** for delivery to the body tissues.

As the heart beats, the sinus venosus contracts and sends nonoxygenated venous blood into the right auricle. Oxygenated blood from the lungs passes into the left auricle via the pulmonary veins and both auricles contract, pumping their contents into the ventricle. Contraction of the ventricle and conus arteriosus next drives the blood to the lungs and the rest of the body. The blood is prevented from flowing back into the heart by means of the semilunar valves located at the base of the truncus arteriosus. Because the amphibian heart is fairly small and does not adequately compare with the structural complexity of a bird or mammal heart, we will next examine a four-chambered mammalian heart.

Mammalian Heart

The mammalian heart is a finely tuned **four-chambered** organ that efficiently separates venous and arterial blood and delivers it to the proper destination. Separate

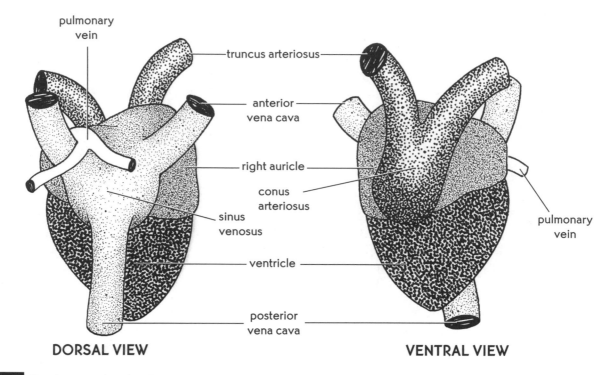

pulmonary
vein

truncus arteriosus

anterior
vena cava

right auricle

conus
arteriosus

sinus
venosus

ventricle

pulmonary
vein

posterior
vena cava

DORSAL VIEW   **VENTRAL VIEW**

FIGURE 9.4 Frog heart, in dorsal and ventral views.

left and **right atria** are again present, and the ventricle is divided by a muscular septum into left and right chambers (Figure 9.5). Obtain a sheep heart, or any large mammalian heart (such as an ox heart), for dissection. The mammalian heart is considerably uniform in structure among vastly different species; most of the structural variation is associated with size. This is a testament to the efficiency with which this organ operates. The sheep heart closely approximates the size and structure of a human heart.

The outer membrane of the heart is termed the **epicardium.** The cavities of the heart are lined with a thin layer of **endocardium,** an epithelial layer similar to the lining of the blood vessels. Between the epicardium and endocardium is the bulk of the heart musculature, the **myocardium.** The orientation of the heart can be identified by first defining the left and right sides. This can be determined by locating the pointed end, or **apex,** which is always associated with the **left ventricle.** The division between the right and left ventricles is indicated superficially by a diagonal furrow containing the coronary blood vessels. The left ventricle is much larger than the **right ventricle** and possesses a thicker muscular wall. Hence, the left ventricle is firm and muscular to the touch, whereas the right ventricle is softer and more flexible. If your specimen has been previously dissected, or is precut, open the heart and proceed to locate the internal structures; if your specimen is intact, make a longitudinal incision from the apex along the left side of the heart. Continue slicing until you reach the left auricle. Now, with a pair of scissors, carefully cut the wall of the auricle, leaving the uppermost tissue

intact. Return to the apex make a similar cut along the right side. You may have to reinsert the cutting instrument to completely slice the **ventricular septum.** Once the cut is complete, you can open the heart completely to reveal the four chambers in longitudinal view.

The left ventricular cavity is larger than the right and the exterior myocardial wall is thicker on the left side. Separating the left ventricle from the superior left atrium is a thin **atrioventricular valve,** often referred to as the **mitral,** or **bicuspid valve.** Opposite to this is the **right atrioventricular valve,** or **tricuspid valve.** Both of these structures regulate the passage of blood from the atria to the respective ventricles. These **A-V valves** attach to the endocardium via fibrous sheaths termed **chordae tendinae** and **papillary muscles.** Tug downward on a few of the cords and note the movement of the valve. The broad muscular ridges lining the ventricles are the **columnae carneae.** Examining the lining of the auricles will reveal a convoluted surface created by the presence of **pectinate muscles.**

Next, identify the large blood vessels which are associated with the heart. Depending upon the preparation and source of your specimen, the vessels may be cropped closely or allowed to extend a short distance. The **aortic arch** exits from the left ventricle and carries blood to the **systemic circuit.** You may find a series of short branches off the aortic arch which separate arterial blood for delivery to the head, arms, and lower body. As the ventricles contract, oxygenated blood from the left ventricle enters the aorta and causes expansion of the smooth muscle of the artery walls. Backflow of blood is prevented by the **aortic semilunar** valve. The right ventricle sends blood to the lungs via the **pulmonary circuit** and the **pulmonary artery.** The right ventricle and the pulmonary artery are separated by the **pulmonary semilunar valve.**

As nonoxygenated venous blood returns from the body tissues, it enters the right atrium via the **superior** and **posterior vena cavae.** Here blood is held between heartbeats until it can flow down the tricuspid valve into the right ventricle. Oxygenated blood returns from the lungs via the **pulmonary veins.** These vessels enter the left atrium just above the position of the **left coronary artery.** The coronary arteries can be found with close inspection of the myocardial walls near the atrioventricular junction.

Frog Arterial System

The arterial system consists of a series of branching vessels that carry blood away from the heart to the capillary beds within most tissues. Arteries carry blood under high pressure; as a consequence they have thick, muscular walls that do not collapse when drained. To facilitate study of the circulatory system, your instructor has provided you with **injected** specimens. This means

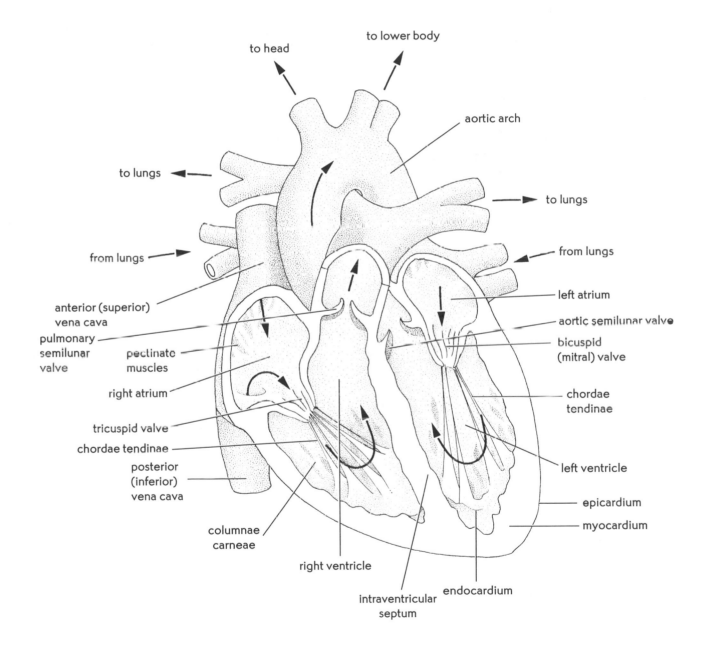

FIGURE 9.5 Longitudinal section of the four-chambered (mammalian) heart.

the supply company which distributes the animal has filled the circulatory system with colored latex to indicate the presence of the blood vessels. In most double-injected specimens, the arteries appear red and the veins blue. If your frog has a set of yellow vessels, these compose the **portal systems** of the liver and kidney. A portal system is one in which the blood is delivered directly to an organ rather than back to the heart after being filtered through a capillary bed. When tracing the path of blood vessels it is helpful to pick an easily found reference point and retrace the pathway to ensure proper identification of the vessel. Locating the vessels and assigning the correct name is similar to reading a road map. The vessels are usually named for the organ or structure with which they unite, and the name frequently changes at the junction of two vessels.

The **conus arteriosus** leaves the heart and immediatley splits into two branches. Each branch, the **truncus**

arteriosus, gives rise to three major arteries (Figure 9.6). The first of these is the **pulmocutaneous arch,** which bifurcates into the **pulmonary arch** (going to the lung) and the **cutaneous artery** (which delivers blood to the skin).

The next artery to come off each truncus arteriosus is the **carotid arch.** It divides into the **external carotid,** which leads to the ventral part of the head and the tongue, and the **internal carotid,** which leads to the dorsal region of the head. The most highly oxygenated blood coming from the heart goes to the carotid arches. Note the small oval swellings, called the **carotid bodies,** at the junctions of the two carotids. The carotid bodies are **chemosensory** structures that are sensitive to changing levels of oxygen in the blood and which also help regulate blood pressure.

The last vessels to come off of each truncus arteriosus are the paired **systemic (aortic) arches.** The two

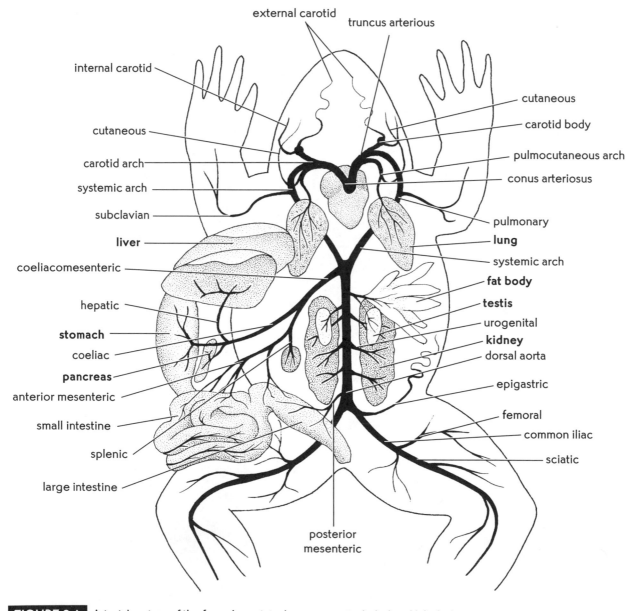

FIGURE 9.6 Arterial system of the frog. Associated organs are included and labeled.

systemic arches bend dorsally around the digestive tract and unite to form the large **dorsal aorta.** Before this union, several blood vessels are given off leading to the skull and vertebral column. The next blood vessel coming off the systemic arch is the **subclavian artery,** which gives off branches to the shoulder region and the forelimbs.

Near the junction of the systemic arches, is the first vessel stemming from the aorta, the **coeliacomesenteric artery.** This vessel gives rise to two branches, the **coeliac** and **anterior mesenteric arteries.** The coeliac then further divides into the **hepatic artery,** going to the liver, and the **gastric arteries,** which run to the stomach. The mesenteric supplies branches to the spleen (**splenic**) and the large and small intestines. The dorsal aorta next gives rise to four to six **urogenital** arteries to the kidneys, gonads, and fat bodies. The **posterior mesenteric artery** arises from the dorsal aorta to supply the colon.

Next, the aorta divides into two common **iliac arteries,** which supply the hindlimbs. As you trace these vessels down the legs, you can observe the **epigastric arteries,** which extend to the urinary bladder, large intestine, and body wall; the **femorals,** which supply the hip and outer part of the thigh; and the **sciatics,** which serve the thigh, shank, and foot.

Frog Venous System

The venous system consists of vessels that carry blood back toward the heart. These vessels have thinner walls but often have a larger diameter than the accompanying artery. The blood flowing through the venous system is under considerably less pressure than that of the arterial system. The head, forelimbs, and skin are drained by a pair of **anterior venae cavae,** which enter the heart by way of the **sinus venosus** (Figure 9.7). Each anterior vena cava receives blood from three

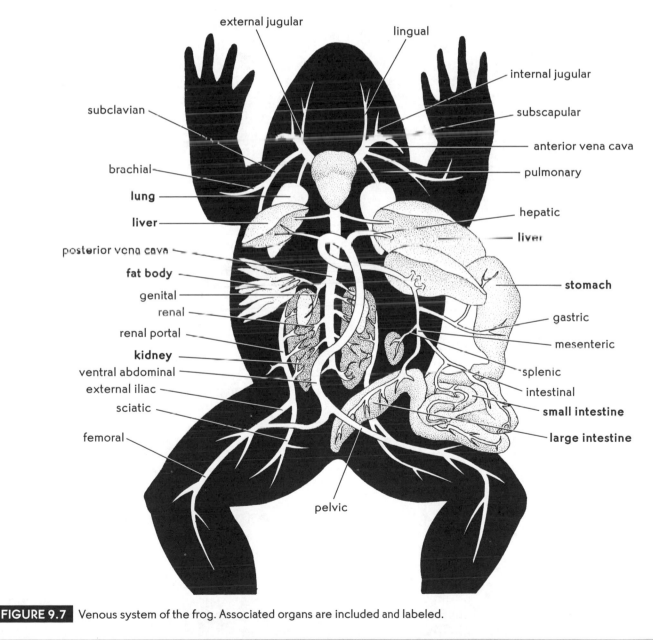

FIGURE 9.7 Venous system of the frog. Associated organs are included and labeled.

veins. The first of these is the **external jugular,** which returns blood from the tongue and floor of the mouth via the **lingual vein,** and from the lower jaw via the **mandibular vein.** The **innominate** veins collect blood from the head by the **internal jugulars** and from the shoulder and back of the forelimbs by way of the **subscapular veins.** The **subclavian veins** collect blood from the forelimbs through the **brachial veins** and from the muscles and skin on the lateral and dorsal part of the head and trunk by means of the **musculocutaneous veins.**

The single **posterior vena cava** enters the sinus venosus posteriorly and collects blood from the testes or ovaries through the **genital vein,** from the kidneys through from four to six pairs of **renal veins,** and from the liver by means of a pair of **hepatic veins.**

The **portal veins** stem from two different portal systems, the **renal** and the **hepatic portal systems.** The renal portal system is found only among lower vertebrates. The renal portal system consists of a **renal portal vein,** which receives blood from the hindlimbs by means of the **sciatic, external iliac,** and **dorsolumbar** vein. The renal portal vein carries blood to the dorsal border of the kidney.

The hepatic portal system consists of a large number of veins that carry blood into the liver from the major digestive organs (the stomach, intestine, spleen, and pancreas). Such blood is heavily laden with recently digested food products absorbed by these organs. As this blood passes through the liver before entering the main circulation, food products are removed and stored, and other substances are added to the blood. In addition, bacteria that may have entered the blood through the digestive tract are phagocytized by specialized (**Kupffer**) cells found in the liver. The main vein draining into the liver is the **hepatic portal vein** which collects blood from the spleen via the **splenic vein,** from the stomach via the **gastric vein,** from the small intestine via the **intestinal vein,** and from the large intestine via the **mesenteric vein.**

UROGENITAL SYSTEM

In lower vertebrates, such as amphibians, the excretory and reproductive organs are usually closely tied. The organs which compose these systems are often united and collectively referred to as the common **urogenital system.**

Urinary System

Several organs compose the urinary system. The most conspicuous organs are paired **kidneys** that lie dorsomedially near the posterior end of the body cavity. Actually, the kidneys are situated behind the parietal peritoneum, a condition termed **retroperitoneal.** The kidneys receive venous blood and filter it through a series of tubules to remove nitrogenous waste. The

form of the waste is associated with the developmental stage of the animal. Like many aquatic animals, the larval tadpole excretes **ammonia,** which is easy to produce and diffuses readily out of the body; ammonia however, carries a rather high toxicity. The adult frog must store its body waste for a time, and hence produces a less toxic compound, **urea.** Urea is more costly to produce metabolically, but eliminates the chance of internal toxicity. The urea is excreted as **urine,** a fluid collected from each kidney and drained by the **ureter.** The ureters empty into the dorsal surface of the cloaca, and the urine must flow across the cloaca to enter the saclike **urinary bladder.** The bladder may be collapsed in your specimen. Look carefully near the posterior edge of the body cavity for a thin, translucent sac.

Running along the ventral side of each kidney, locate a discolored strip of tissue. These are the **adrenal glands,** which secrete adrenaline (epinephrine) into the blood. This hormone is released in response to stress; it increases the heart rate and blood pressure, and raises blood glucose levels. You may have to look carefully for these glands or observe another specimen where the glands are better developed.

Female Reproductive System

The reproductive organs vary greatly in size and development in female frogs relative to the stage of the breeding cycle. The **ovaries** in mature female frogs are quite conspicuous, and can occupy much of the body cavity when they are actively producing eggs (usually the fall and winter). After ovulation in the spring, the ovaries shrink approximately to the size of the urinary bladder. A mature female with eggs will possess a large black ovary, while immature females or females at the end of their breeding cycle will display a small cream colored ovary. The development of the paired **fat bodies** is also closely tied to the reproductive cycle. The fat bodies are metabolized during egg production and are often quite reduced when the ovary is swollen.

If your specimen has a pair of large, well-developed ovaries, remove them to expose the remaining urogenital organs (Figure 9.8). Embedded in the parietal peritoneum, locate the **oviduct,** a highly looped organ which extends from the ovary forward to the anterior end of the body cavity. At the proximal end of the oviduct is a funnel-shaped **ostium.** When eggs are released from the ovary, they first pass into the body cavity and are carried to the ostium by currents created by movement of the cilia of the coelomic epithelium. The posterior ends of the oviducts form enlarged **uteri,** whose entrance to the **cloaca** is on the dorsal surface near the entrance of the ureters.

Male Reproductive System

In a male specimen the **fat bodies** are usually much more conspicuous than in most females. If the fat

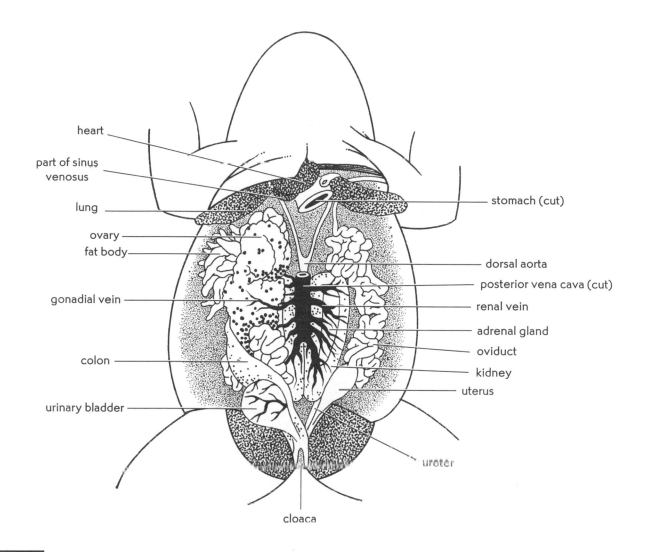

heart

part of sinus venosus

lung

ovary

fat body

gonadial vein

colon

urinary bladder

stomach (cut)

dorsal aorta

posterior vena cava (cut)

renal vein

adrenal gland

oviduct

kidney

uterus

ureter

cloaca

FIGURE 9.8 Ventral view of the female frog urogenital system. An ovary and fat body have been removed on the right side of the drawing.

bodies are so well-developed that they conceal any underlying organs, remove them, noting their oily texture. Males will store excess energy in the fat bodies to be metabolized during hibernation. On the ventral surface of the kidneys, locate two oval, yellowish **testes** (Figure 9.9). Gently lift up one of the testes with a pair of forceps and note the fibers that pull up from the

surface of the kidney. These are the **vasa efferentia,** a series of thread-like tubes that carry sperm to the kidneys. Sperm are then carried from the kidneys to the cloaca via the ureters, which also serve as the urinary ducts. Your specimen may show a vestigial oviduct (or **mesonephric duct**) embedded dorsally into the parietal peritoneum.

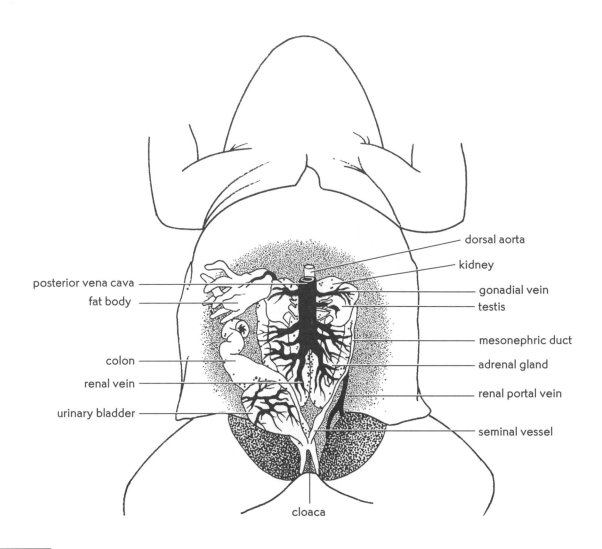

posterior vena cava

fat body

colon

renal vein

urinary bladder

dorsal aorta

kidney

gonadial vein

testis

mesonephric duct

adrenal gland

renal portal vein

seminal vessel

cloaca

FIGURE 9.9 Ventral view of the male frog urogenital system. One fat body has been removed from the right side of the drawing.

EXERCISE 10

Sensory Systems

Sensory reception and processing of environmental information is essential if organisms are to adapt to the variability of environmental factors. Three sets of experiments are outlined in this exercise. Two of these deal with primitive animal species which show limited sensory development, and the third examines the human senses. Your laboratory instructor will divide you into experimental groups, each of which will perform one of the invertebrate experiments and all of which will complete the human exercise. Read the statistical theory statement in Appendix I and write up a laboratory report, based on the experimental results, and statistics generated, which you obtained in one of the experiments. Use the instructions provided by the laboratory instructor or those found in Appendix II to format your report.

GEOTAXIS IN SNAILS

Terrestrial snails (*Helix* and other genera) show a natural inclination to creep upward when they encounter a slope. In nature, this sloped surface is often the stem of a plant or perhaps a rock surface. In the laboratory we can simulate this surface and note the animal's reaction. If, while a snail is actively crawling, the surface is rotated smoothly so that the snail is now placed in a horizontal position, the animal will reorient itself and, after a latent period, will continue its upward journey.

This sensory response can be traced to action of the statocysts, the effect of air currents on the animal's sensory tentacles, or the altered pull upon its body musculature caused by the new position.

To perform this experiment, you will need to obtain the correct apparatus, as shown in Figure 10.1. This is a fairly simple device utilizing a rotating piece of plexiglass that may be rotated around a central point. The center is fastened to an underlying hinged board which can be tilted to a number of set angles relative to the horizontal. To ensure isolation of interfering forces, avoid direct illumination.

Obtain a living specimen of *Helix* or another suitable species such as *Littorina*. *Littorina* is an aquatic species that routinely encounters tall stalks of *Spartina* grass in its environment. Place the snail on the plexiglass, in a near vertical position, and observe its behavior. Handle the snail gently to prevent a shock reaction. Note its movements and the time it took the animal to orient itself and begin moving.

Rotate the plexiglass 90°, so that the snail is crawling horizontally, and accurately measure the time which elapses before reorientation occurs. Activity is signaled by the first movements of the head, followed by foot action. Determine the length of this time period by making several trials on the same snail and calculating the mean. Next, try rotating the board first to the right, then at the same angle to the left. Note whether

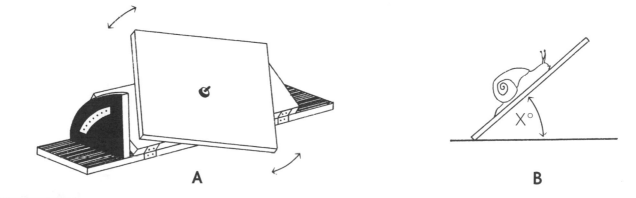

FIGURE 10.1 A) a simple turntable with adjustable slope; B) "steepness" expressed by angle (X); "slope" by tangent of (X).

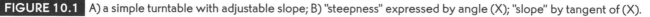

the animal responds as readily to one side as to the other. Repeat these measurements in exactly the same order for 3 more specimens. You can now compare means between replicates and utilize a *t-test* to determine if there is a statistical difference (see Appendix I). Repeat the measurements at different angles, decreasing the vertical steepness each time. Try attaching a small weight (such as a fishing sinker) with some fine fishing line and carefully tape it to the shell. Observe the affect the weight has on reorientation time.

Experimental Protocol

The steepness of the plate may be expressed by the angle (X). The slope is, technically, the tangent of (X), but it is sufficient to simply use the angle. With the snail crawling upwards on the vertical plate, the plate is rotated 90° and the reorientation time or latent period measured. Repeat this with the plate sloping at decreasingly smaller angles (X = 80°, 60°, 45°, 30°, and 0°) and measure the latent period of the snail's response at each slope. On a piece of graph paper, plot the latent period against the angle to show the relation between steepness of slope and repsonse speed. Include the graph as a figure in your lab report.

PLANARIAN RESPONSES TO LIGHT

The ability to orient the body in response to differing directions and intensities is termed **phototactism**. This is a sensory capacity common to many animals and plants, but the complexity of the response and the mechanisms involved vary greatly among different levels of advancement.

Responses to Directional Illumination

Using a focused source of light and a dissecting microscope, observe the movements of a freshwater planarian (*Dugesia*) in a small dish of culture water. Begin with the light source diffusing light laterally across the medium. Now switch the light source to a position 90° to the dish and observe the animal's reaction. *Dugesia* lives on the undersurface of rocks, logs, and leaves in aquatic environments where it actively scavenges for detritus. Thus, the animal is **negatively phototactic**; it avoids direct illumination and will quickly seek shelter from high intensity light. Repeat this procedure several times and note if the initial direction of the turn is consistent. If not, generate a hypothesis and present this in your lab report.

Shock Reactions

If a planarian encounters a sharp boundary between a dark and a strongly lit region, it may exhibit "shock behavior," a reaction which may prevent it from crossing the boundary. This type of response can be easily demonstrated by preparing several glass plates with distinctly dark and light areas. Place a watch glass containing a planarian over the surface of the plate and illuminate the whole apparatus evenly. Note the reactions as the planarian occasionally starts to leave the shadow and reacts to the light region. Use a series of plates darkened to different intensities to estimate how much intensity difference is necessary to produce a shock reaction. If you wish to quantify the light intensity values of the glass plates, use the light meter of any 35 mm camera.

Background Selection

Planarians will choose a particular surface to come to rest on, based on differences in light intensity. In order to avoid directional illumination, and the influence it might have on the animal, place the animal in a watch glass with one half of the bottom and the sides covered with black tape or paint. Cover the remaining half of the bottom with white paper or paint. Cut out a circular piece of black paper in such a way that it will form a small lip across the top of the dish. This is to ensure that directional light does not enter the dish. Now place these dishes in a darkroom or in a blackened box to shield them from further lateral illumination. Several feet above the dishes place a source of low intensity light, being sure to avoid lights which will generate excess heat. Place several planarians in the center of the dish and leave them undisturbed for 15 minutes or until they have ceased to move about actively. Count the number that have come to rest on the dark side of the dishes and compare with the number on the white side. Determine the percentages of specimens resting on the black versus white surface. You can follow the courses taken by the animals at the end of the experiment (after removing the planarians) by adding a suspension carmine powder to the water, and gently shaking the dish. The particles stick to the mucous trails left by the planarians. Is there a difference in the amount of wandering on the black versus the white surface? Answer this question and present your data in your lab report.

REFLEXES, CUTANEOUS RECEPTORS, VISION, AND TASTE IN MAN

The somatic senses provide the brain with information about the condition and position of the body. Various kinds of pain and pressure receptors are associated with all internal organs. Certain of the internal receptors important for feedback regulation, for example at joints and in muscles, are termed proprioceptors. There is some overlap, for example the **cutaneous receptors** provide information about the body as well as information about the environment.

The special senses are designed to provide the brain with information about the environment. Each system contains specialized neurons called receptor cells that transduces the energy of the environmental stimulus into a neural impulse. The eye transduces light energy, the ears transduce sound, and the tongue and nose (nasal epithelium) transduce chemicals into generator potentials and then into action potentials (impulse). Each special sense is associated with a specific cranial nerve or nerves. Stimulation of olfactory cells in the olfactory epithelium in the nasal cavity lead to impulses all of which enter the brain via the olfactory nerve (cranial nerve I). All auditory input to the brain is via the auditory nerve (cranial nerve VIII). Though several cranial nerves provide motor neurons to regulate movement of the eye (cranial nerves III, IV and VI), all visual sensory input to the brain is via the optic nerve (cranial number II). Impulses from the taste buds enter the brain through the facial nerve (cranial nerve VII).

Reflexes

A reflex is an involuntary response to a stimulus. A reflex arc or pathway involves most often three neurons: the sensory or afferent neuron, usually an interneuron, and a motor or efferent neuron (Figure 10.2). The receptor cell or sensory neuron transduces the stimulus to a nervous impulse and transmits it to the central nervous system. The interneuron makes connections to the motor neuron and to the rest of the nervous system. The motor neuron transmits the impulse to the **effector organ** — usually a muscle.

Question: Why is a three neuron spinal reflex so much quicker than a response that must be recognized by the brain?

Visceral and somatic reflexes

The **visceral reflexes** are involuntary responses of the viscera which regulate responses such as breathing, heart activity, and blood pressure.

The **somatic reflexes** are involuntary responses of the skeletal muscle system and include the patellar response, muscle spindles, pupillary response, and nystagmus. Some of these are termed proprioceptors.

Proprioceptors

Mechano-type receptors located in the joints and in the muscles convey information about the position and location of body parts to the brain. These include spindles in the muscle fibers, Golgi tendon organs (sense stretching) and ligament receptors (joint receptors). Some proprioceptors operate more effectively with supplemental input from the visual system.

Visual dependent proprioceptors

a. Close your eyes and have your partner see how long you can stand on one leg with your arms extended (report the units in seconds)._____

b. Repeat step a. with your eyes open. How important is visual input to the proprioceptors involved in balance?_____

Visual independent proprioceptors

a. Close your eyes and extend your arms fully to the sides. Bend your elbows and bring your four fingertips together. Your partner can measure how close you were in case you missed. Try several times. How close were you?_____

b. Can you do appreciably better with your eyes open?

_____

The response of some somatic reflexes involving skeletal muscles are useful in diagnosing injury to various parts of the nervous system. Observation and testing of some of these reflexes are useful in elucidating the function of the nervous system.

Pupillary Reflexes

The iris muscles of the eye control the amount of light entering the eye. The pupil **dilates** automatically in dim light and **constricts** in bright light. Pigment in the iris is responsible for eye color.

a. Shine a pocket flashlight beam into the eye of your partner and observe the constriction of the pupil. Observe the dilation when you remove the light beam. Which cranial nerve is responsible?_____

_____

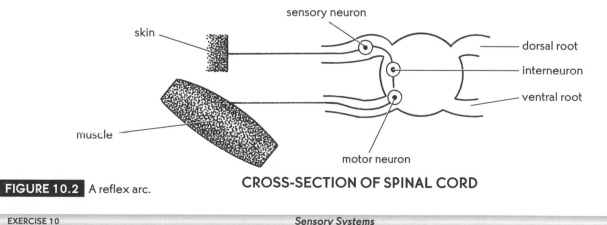

FIGURE 10.2 A reflex arc.

CROSS-SECTION OF SPINAL CORD

b. Hold the edge of your hand along the nose of your partner so that the beam of light only impinges on the one eye. Does the pupil of the other eye constrict? _____ If so, why? _____

_____

_____

_____

_____

c. Keep the pupil of your partner's eye constricted by maintaining the light beam on the eye. Reach behind him and gently pinch him on the nape of the neck. The pinch should be sufficient to produce a "chill" effect. Note the brief dilation of the pupil even though the eye is still illuminated. What is the role of the autonomic nervous system in this chill reflex? _____

_____

_____

_____

_____

_____

_____

_____

Patellar Reflexes and Facilitation

The patellar tendon attaches to the tibia and can be located just below the knee bone or patella.

a. Have your partner sit on the edge of a table with the legs hanging free. Strike the patellar tendon, located just below the knee, with a rubber reflex hammer. The lower leg will kick out (jerk). What muscles are responding? _____

b. If there is no response have your partner look away and think of something else. Cerebral inhibition can block the response.

c. Have your partner push his palms together with all his strength as you strike the patellar tendon. Notice the much stronger reflex response. This phenomenon is termed facilitation and is caused by interactions at the spinal level in which the activity caused by arm muscle contraction is transmitted to the motor neuron, then to the leg.

Cutaneous Receptors

Cutaneous receptors are sensory nerve endings in the dermis. There are basically four types: pain, pressure, hot and cold. Pain endings are simple dendritic fibers not associated with any structure. The various kinds of pressure sensitive endings and the temperature receptors (including touch) are associated with end bulb structures made of dermal cells.

Discrimination

Prepare a set of dividers, such as used by draftsmen, by slightly blunting the pointed tips. The subject should feel two points, but one does not wish to draw blood.

a. Close your eyes. Your partner should spread the points far enough apart so that when touched to the back of the hand both tips are felt. Bring the points closer together and again touch both tips. Repeat until you can only distinguish one tip. Vary between one and two tips at the final stage to reduce the error.

b. Measure the minimum distance between the two points when they can still be perceived as two points. Record the results on the following chart and repeat the test on the fingertip and the forearm.

Part tested	mm separation
Back of hand	_____
Fingertip	_____
Forearm	_____

Question: What conclusions may be drawn concerning the density of touch receptors in the various parts of the body?

Thermoreceptors

There are two specialized nerve endings in the dermis of the skin for the perception of temperature. Krause's end bulb responds specifically to cold and is more sensitive than the heat receptor. The heat receptor is more difficult to distinguish from touch at lower temperature and from pain at higher temperature.

Temperature probes are easily made from 5 or 10 ml plastic syringes by fitting a length of heavy gage aluminum or copper wire through the needle end of the syringe (Figure 10.3). The wire should fit snugly so that it stays in place and the exposed tip be sanded round.

a. Fill one syringe with ice water and chopped ice, the other with hot water and plug both with a rubber stopper.

b. Mark off a square of 1 × 1 cm with a black felt tip pen. Have your partner touch the skin lightly in the square and if it feels cold make a blue dot (blue pen). If you only felt it make no mark. Find as many cold receptors as possible in the square.

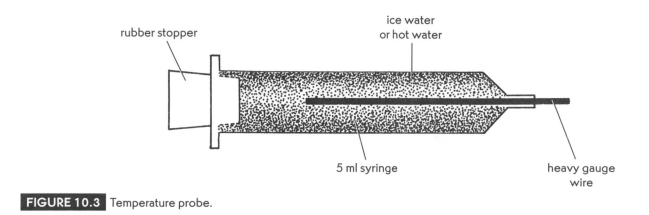

rubber stopper

ice water
or hot water

5 ml syringe

heavy gauge
wire

FIGURE 10.3 Temperature probe.

c. Repeat with the hot probe and find as many warm receptors as possible.

d. Tally the cold and hot receptors. Which are more common in the marked off area? _____

Cold receptors _____ per cm^2

Heat receptors _____ per cm^2

Vision

Humans are more dependent on their eyes than on any other sensory organ. In man, the eye is large and very well-developed. The spherical shape of the eye depends upon the pressure of the fluids within the eyeball. The focusing power of the eye is based on the bending of light as it passes through the curvature of the cornea and the lens. Clear vision depends on a sharp image being focused on receptors capable of discriminating intensity and wavelength. Adjustments to the focus, to compensate for the distance of an object from the eye, is made through a change in the curvature of the lens, a process called **accommodation**. The photoreceptors in the eye are located in the retina and are of two types: rods and cones. The rods operate in dim light and contain only one kind of visual pigment. The cones operate at high light intensity and are concentrated in the fovea, the area of most acute vision (where we normally focus). Cones are associated with color vision. Transduction of light occurs when a photon is absorbed by the visual pigment in the rod or cone, and the vitamin A portion of the retinalaldehyde undergoes a structural change. The pigment is thereby bleached and requires a few minutes to be regenerated. The structural change in the visual pigment leads to membrane depolarization and this contributes to a generator potential.

Visual acuity

Measure the focusing capacity of the lens plus cornea system of the eye.

a. Stand exactly 20 feet from a Snellen Eye Chart and cover one eye with a 3 × 5 card.

b. Read all of the lines you can see clearly without squinting or straining. Note the line number and determine the d/D. d is the distance (20 ft) at which the subject is reading the line. D is the line number, which is the distance at which a normal eye could read that same line. For example a slightly myopic eye (nearsighted) would be 20/100, meaning the smallest line the subject can read at 20 ft is line 100, which is what a normal eye could read at 100 ft.

c. Repeat with the other eye and then try it with or without glasses if you wear them.

	d/D without glasses	d/D with glasses
Right eye	_____	_____
Left eye	_____	_____

Accommodation

This is the process of changing the shape of the lens in order to control the bending of light (called refraction) to focus on the retina. As an object is moved closer to the eye the ciliary muscle contracts and releases the tension on the edges of the lens. This allows the lens to become more curved (thicker in the middle) due to its natural elasticity. As an object is moved closer to the eye a point is reached where its image becomes blurred. This is the near point of accommodation. Estimate the near point of accommodation using the following procedure.

a. Hold a printed page at a distance at which it can be read, then gradually bring the page toward the eye until the print becomes blurred. Gradually move the page away until the print just becomes readable. This is the near point of accommodation.

Near point of left eye = _____ cm

Near point of right eye = _____ cm

b. The elasticity of the lens diminishes with age and the near point of accommodation becomes greater. The following table shows the relationship.

Age (years)	Near point (cm)
10	7.5
20	9.0
30	11.5
40	17.0
50	52.5
60	83.0

Draw a graph of the age versus the near point and determine the age of your eyes.

Question: How do myopia and hyperopia affect the near point of accommodation? Did you take this into account when you were determining the age of your eyes?

Demonstration of the blind spot

The area on the retina where the neurons from the rods and cones pass into the optic nerve is called the optic disc or blind spot. There are no photosensitive cells (rods or cones) in this area, which means that an image falling on this spot would not be perceived.

a. The figure below shows a cross and a circle about 70 mm apart. Copy them to a 3 × 5 card with the cross to the left side and use them to demonstrate the blind spot.

+   O

b. Hold the card about 50 cm in front of your face with your nose pointing halfway between the cross and the circle. Close your right eye and concentrate your left eye on the circle. Slowly move the card toward your face. The cross should disappear. If it doesn't, you were not focusing on the circle. Try it again.

c. The cross should reappear again as you continue to move the card closer to your face.

Question: Why haven't you ever noticed a blind spot in your vision before?

Colored afterimages

These are the result of selectively bleaching out one of the three colors of cone pigments. People who stare for long periods at bright green computer monitors find a red tinged world (red is the complement to green) when they look away. Set up the following experiment.

Place a small square of solid blue paper or cardboard on a sheet of white paper, and place it under a very bright light.

a. The test subject should stare at the blue square for about one minute, then shift his gaze to a blank piece of white paper. Start timing when the subject shifts his gaze to the blank paper.

b. What color was the afterimage? _____

c. How long did the afterimage last? _____

_____

While staring at the blue square, the cones containing the pigment sensitive to blue light were partially bleached — becoming less sensitive. The cones containing red and green sensitive pigments were unaffected. When you shifted your gaze to the white surface, then white light entered the eye (which contains all colors), but only the red and green receptors (cones) were fully stimulated. The brain interpreted this combination as yellow, as though the eye were receiving yellow light. The brain perceives the afterimage as yellow until the blue receptors fully recover and the afterimage fades.

Chemoreceptors

Chemoreceptor cells respond to chemicals in and on the body. Of primary importance are the olfactory receptors located in the olfactory epithelium that line the roof of the nasal cavity and the taste receptors located in taste buds on the tongue. The olfactory receptors are in general more sensitive and whereas the tongue is limited to about four basic tastes, there are literally thousands of different odors that can be distinguished by the nose.

Map the location of taste receptors on the tongue

Map out the locations on the tongue that respond to the four basic tastes. There will be some overlap of taste buds, and also some taste buds respond to more than one taste. Test the numbered areas of the tongue shown in Figure 10.4.

Use the following procedure to determine the taste responses, and use dropper bottles containing the following solutions:

5.0% sucrose (sweet)   0.1% quinine sulfate (bitter)

10% salt (salty)   1.0% acetic acid (sour)

a. Open your mouth and do not stick out your tongue. Have your partner place one drop on spot #1. Rinse

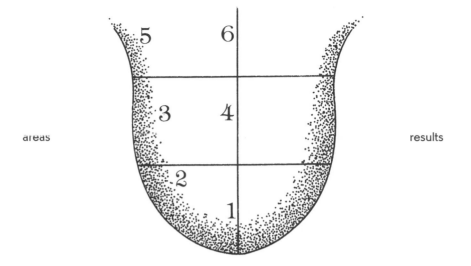

areas   results

FIGURE 10.4 Upper surface of the tongue.

your mouth with water, record your response, and repeat for all other spots.

b. Repeat for the other three solutions.

c. Write in the symbols **sw** (sweet), **sa** (salt), **bi** (bitter), and **so** (sour) to correspond to the numbers on the other side of the drawing of the tongue. Map the areas with pencil shading or colored pencils.

Cardiovascular Function: Heartbeat and Blood Pressure

This laboratory will be done using two major experimental designs. The first will deal with the temperature coefficient (Q_{10}) of the heart rate and respiratory movements of the waterflea *Daphnia*, and the second will look at the human heartbeat and blood pressure in relation to the effects of posture and exercise. The student groups will be partitioned so that each group will be working on either Q_{10} effects on *Daphnia* or heartbeat and blood pressure effects on man. As in exercise 10, students will be responsible for material in both exercises so you should be sure to observe the other exercises in progress.

Q_{10} OF HEART RATE AND RESPIRATORY MOVEMENTS IN *DAPHNIA*

The ambient temperature often has a profound effect upon the activity pattern and metabolic rates of animals. This is especially true of ectothermic animals, like the lower vertebrates and all invertebrates. The degree to which a metabolic rate process is affected by the temperature is most referred to as the Q_{10} value, or the **temperature coefficient**. The Q_{10} is determined by the ratio of the metabolic rate at a given temperature to the rate at a temperature 10°C lower. The Q_{10} varies greatly over different parts of the temperature range, even with similar biological systems. Since most vital life processes become increasingly slower as temperatures near 0°C, the Q_{10} for the lower ranges of temperature is often high while, as the upper limits of thermal tolerance are approached, the Q_{10} drops. This implies that the Q_{10} for any animal is not a constant and the Q_{10}'s of different processes in the same animal may not be the same.

The temperature coefficient (Q_{10}) of a process is the ratio of its rate, or velocity constant, at a given temperature to its rate at a temperature 10°C lower. Thus the Q_{10} for a process of rate K_t at temperature (t) and rate $K_{(t + 10)}$ is:

$$(1) \quad Q_{10} = \frac{K_{(t + 10)}}{K_t}$$

It is not necessary to use temperatures 10°C apart since, if the rates at any two temperatures (K_1 at temperature t_1 and K_2 at temperature t_2) are known, then van't Hoff's equation applies:

$$(2) \quad Q_{10} = \frac{K_1}{K_2}^{\left(\frac{10}{t_1 - t_2}\right)}$$

This equation is more easily handled in the form (3):

$$(3) \quad \text{Log } Q_{10} = \frac{10}{t_1 - t_2} \bullet \text{Log} \left(\frac{K_1}{K_2}\right)$$

Experimental Protocol

To determine the temperature coefficients (Q_{10}) of two processes within the same animal, observe and record the rates of both the **heartbeat** and the **respiratory movements** in the freshwater crustacean *Daphnia* at a series of temperatures from 0°C to 30°C. At small temperature intervals measure the time for 10 beats (using a stopwatch). Take at least three readings for each temperature, and determine the mean. The reciprocal of the time for 10 beats is a convenient measure of the rate. After completing a temperature series, check at some intermediate temperature to see if the effect is reversible. If it is not, the extremes of temperature may have injured the animal and invalidated the results. Plot mean rate against temperature in degrees Centigrade on both arithmetic and semilogarithmic paper. Note the form of the curve and its differences on the two types of graph paper.

To obtain readings at a series of temperatures, place the animal in a vessel of water at 0°C and allow it to warm up to room temperature. To elevate the water temperature above room temperature, have a supply of hot water handy. Allow the animal time to adapt at each temperature interval.

In formulating the discussion of your lab report, consider the state of acclimation of the *Daphnia*. Ask your lab instructor if the *Daphnia* were reared or maintained at different temperatures; if so, is there a difference in the Q_{10} or in the upper and lower limits of normal activity?

HEARTBEAT AND BLOOD PRESSURE: EFFECT OF POSTURE AND EXERCISE

Each time the ventricles of the heart contract (**systole**), a specific quantity of blood (**stroke volume**) is forced into the arteries. The walls of the major arteries are composed of a layer of smooth muscle and a layer of elastic fibers, and the force of the blood in the arteries causes them to expand. The force that causes this expansion is the **systolic pressure** and a wave of expansion travels down the major arteries with each contraction of the ventricle. This wave can be felt as the **pulse** in most of the larger arteries. The pulse therefore is a measure of the rate of heartbeat. When the ventricles relax (**diastole**) the expanded arteries contract due to the contractive force of the elastic fibers in their walls. This force of the elastic fiber contraction causes a **diastolic pressure**. The blood pressure is always measured as systolic/diastolic. The average systolic pressure is 130 mm Hg and the average diastolic pressure is 80 mm Hg. The mm Hg refers to mm of mercury pressure above the 760 mm atmospheric pressure.

The blood pressure is greatest in the large arteries close to the heart and drops off in the smaller arteries farther away from the heart. It is usually measured in the branchial artery of the upper arm. The blood pressure is generally higher in older people and lower in people in good physical condition. It is under control of the autonomic nervous system and therefore elevated by certain drugs and behavior. Caffeine in coffee, nicotine in cigarettes, and stress all elevate the blood pressure. A consistently high blood pressure, especially a diastolic pressure above 100, indicates **hypertension.**

The baseline heart rate is not dependent on nervous stimulation from the nervous system, but rather from intrinsic modified heart muscle tissue called **pacemakers.** The rate of heartbeat can be increased or decreased by nerves of the autonomic system. Exercise increases the heartbeat because it activates the sympathetic system (part of the autonomic nervous system). When we lie down the heart slows because acetylcholine is released from the vagus (part of the parasympathetic system). Excitement speeds up the heart via the hormone epinephrine released from the sympathetic system.

Effects of Posture on Heartbeat

This experiment can be done in conjunction with the effect of posture on blood pressure. Measure the heartbeat of your partner by palpation (feeling) with the middle finger of the radial artery in the wrist. If your partner is busy measuring your blood pressure, then measure your own pulse rate (heartbeat) by palpating your carotid artery in the neck. Count the pulse for 15 sec and multiply by 4 to get beats/min.

Determine the average of three pulse readings of the subject in the prone position lying on the laboratory table. Have the subject quickly stand up, and immediately count the pulse. Note the extent of the increase and determine the average increase.

Average prone pulse	Average after standing	Average after 5 min

Immediately after standing, gravity momentarily retards the return of blood to the heart, and the blood pressure declines. The autonomic system quickly responds and speeds up the heartbeat in order to maintain constant cardiac output. Within minutes pressure and pulse rates return to normal.

Procedure to Measure Blood Pressure

This procedure will be demonstrated by the lab instructor:

1 By means of the Velcro overlap, secure the cuff of the sphygmomanometer around the upper arm of your partner.

2 Face your partner and rest his/her arm on the bench or on your hip with the cuff at about the level of the heart.

3 Place the membrane of the stethoscope in the crook of the elbow just below the cuff.

4 Inflate the cuff rapidly to about 180 mm Hg. No sound should be heard because the branchial artery has been completely occluded.

5 Keeping your eye on the pressure gauge, immediately release the pressure at the rate of about 5 mm per second.

6 The first sound you hear is a clicking sound as blood spurts through the constricted opening in the branchial artery. This is an estimation of the systolic pressure.

7 As the pressure falls the sound becomes louder and the characteristic lub-dub heartbeat sound is heard.

8 As the sphygmomanometer pressure continues to fall the sound becomes muffled and disappears as the branchial artery is completely opened. The disappearance of sound is an estimation of the diastolic pressure.

9 Release the rest of the pressure to zero as quickly as possible to relieve the discomfort on the arm of your partner.

Take three readings of blood pressure of your partner in the prone position and three in the standing position; now determine the mean and record this value. Your partner may be simultaneously taking his/her pulse. Allow the subject 2–3 minutes of rest between each measurement.

	Systolic pressure	Diastolic pressure
Reclining position	_____	_____
Standing position	_____	_____

Question: Why is the systolic pressure greater for a brief period in the reclining position than in the standing position?

Effects of Exercise on Blood Pressure

The extent of the response of the cardiovascular system to exercise stress in humans is directly related to the conditioning of the body and its age. In general a younger person, a person in better health, and/or a person in better physical condition will experience a smaller increase in pulse rate and blood pressure after exercise than older people or those in poorer physical condition. It is normal for some increase in pulse and blood pressure to occur with exercise even in a person in top condition. Normally, only the systolic pressure increases with exercise and not the diastolic. Elevated diastolic pressure when at rest indicates increased back pressure and more strain on the heart and is a strong indicator of hypertension. Elevation of diastolic pressure during exercise is not normal and should be further investigated. In healthy individuals the increased pulse and systolic pressure in response to exercise will return to normal more rapidly than in individuals who are not in as good a physical condition. In well-conditioned individuals it will require more exercise in order to elevate pulse and blood pressure.

Determine the normal resting (prone) and standing pulse and blood pressure of your partner. Leave the deflated cuff in place. Next have him perform a standard exercise. Use a sturdy chair of standard height (45 cm). Place one foot on the chair, slowly raise the body, without using supports, until both feet are together — then lower the left foot and body to the original position. Repeat this exercise 5 times at 3 second intervals. Immediately have him take his own pulse readings every 15 secs for two minutes while you take his blood pressure every minute. If the pulse and blood pressure have not returned to normal in the two minute period, you can estimate the time it would take to return to normal based on your data and the line you plot on your graph (see below).

	Pulse rate	Systolic pressure	Diastolic pressure
normal	_____	_____	_____
15 sec	_____		
30 sec	_____		
45 sec	_____		
60 sec	_____	_____	_____
75 sec	_____		
90 sec	_____		
105 sec	_____		
120 sec	_____	_____	_____

Time to return to normal pulse rate: _____

Time to return to normal blood pressure: _____

Construct a Graph

Using the above data, construct a graph with two Y axes. Use one Y axis for pulse and the other for the blood pressure units. The X axis will of course be the Time scale. You must of course use different symbols to avoid confusion, for example a circle for pulse, a square for diastolic, and a triangle for systolic pressure. Alternatively, one could draw the lines in different colors or use dashed and solid lines.

Question: What are the roles of the sympathetic and parasympathetic nervous system in the regulation of pulse rate and blood pressure. How does acetylcholine and norepinephrine (and epinephrine) affect the heart and blood vessels?

EXERCISE 12

Metabolic Processes

The following exercise is composed of a series of experiments analyzing crucial metabolic processes typical of many animals. Your instructor may choose to split the class into groups, as was done with the previous physiological experiments, or you may select only one of the experiments depending on time constraints or the availability of the experimental apparatus.

OXYGEN CONSUMPTION

Oxygen consumption in an animal is a common method for the indirect measurement of an animal's metabolic rate. The direct method to measure the metabolic rate is to measure total heat production, which can be done by using controlled temperature baths. The direct method is not practical for laboratory purposes, hence, we will use an alternative method. The metabolic rate or oxygen consumption is related to the activity of the animal and measures the pace at which the body is running. The rate is influenced by environmental temperature and the diet of the animal.

The metabolic rate is measured in calories/gram body weight and the oxygen consumption in ml oxygen/hr. The relation between oxygen consumption and metabolic rate can be estimated by examining the values obtained from the oxidation of food.

Carbohydrates	Lipids	Proteins	Units
4.2	9.5	4.2	cal/g
0.82	2.0	0.92	ml oxygen/g animal
5.12	4.75	4.68	cal/ml of oxygen consumed

The oxidation of an average combination of carbohydrates, lipids, and proteins will produce a value of 4.8 cal/ml O_2. Therefore if you multiply the ml of oxygen consumed per hr by 4.8 you will have a good estimate of the metabolic rate.

Estimated metabolic rate = ml of O_2 consumed/hr $\times$ 4.8 cal/ml O_2

Procedure

We will use a liter glass jar as a respirometer to measure oxygen consumption in a frog. Note the various parts as labeled in Figure 12.1. Place a fresh bag of Indicator Soda Lime (wrapped in cheese cloth) in the bottom of a wide mouthed glass jar. The soda lime must be fresh and kept dry, or it will not perform its purpose of absorbing the carbon dioxide exhaled by the animal. Place the U-shaped piece of hardware cloth over the soda lime bag to form a platform for the animal to sit upon. Carefully push the 18 gage needle and the 1 ml pipette through the two openings in the large rubber stopper. The pipette is pushed through much easier if moistened with soapy water. Use the soapy water to blow bubbles, then suck the bubbles into the pipette. The bubbles will form a marker that will move as the animal consumes oxygen.

Place the animal on the screen and carefully twist the stopper in place to form an air tight seal. Pull the plunger of a 5 ml plastic syringe to the 5 ml mark and carefully twist it into the plastic base of the needle to form an air tight seal. If everything is properly sealed, pushing on the plunger will cause the soap bubble in the pipette to move away from the jar. If no movement of the soap bubbles occurs there is a leak. If the bubbles are lost, start again.

If the respirometer is not leaking, then proceed to make a run as follows. Note the location of a particular bubble or group of bubbles in the pipette. As the animal consumes oxygen it is also giving off carbon dioxide. The carbon dioxide is absorbed in the soda lime and not added to the jar volume, but the oxygen used up by the animal is subtracted from the total jar volume. As the oxygen is consumed the soap bubbles will move towards the jar. Note the time and the reading of the syringe. Gradually add the air from the syringe into the jar before your marker bubbles are lost. When the syringe is almost empty, bring the marker bubble to your starting location and note the time and the reading on the syringe. From this determine the ml of oxygen consumed per X minutes.

You and your partner should set up a respirometer as explained previously. In order to determine the oxygen consumption, the soda lime bag must be in place when the experiment is run. When using frogs, do not let the frog become too dry. Desiccation will impair its respiration rate, because amphibians rely heavily on cutaneous respiration. Periodically during the lab place the frog in a jar with pond water (not tap water) and allow it to drain before placing the animal back in the respirometer.

Each pair of students should determine the oxygen consumption and metabolic rate of a lower vertebrate with a slow metabolic rate, such as a frog, and a higher vertebrate with a fast metabolic rate, such as a small rodent. Different student groups should try smaller versus larger animals. Put your results on the board and complete the chart below.

1. Smaller animals of the same class (frog versus frog; rodent versus rodent), should have a higher consumption **rate** but probably a lower total oxygen consumption. Rate is based on oxygen consumed per gram animal. Therefore, weigh the animal that you use.

2. Rodents should have a higher oxygen consumption rate than frogs. To avoid the circus of frogs jumping all over the class, be careful when handling the animals and when returning them to their container make sure the lid is secured.

	ml O_2 consumed per ___ min	consumed per hour	consumed per hour per gram
Frogs:			
Small	_____	_____	_____
Large	_____	_____	_____
Rodents:			
Small	_____	_____	_____
Large	_____	_____	_____

Using the value of 4.8 cal/ml oxygen and the value of ml oxygen consumed/hr determine the metabolic rates of these animals (e.g., multiply 4.8 cal/ml by the obtained values for oxygen consumed per hour).

Metabolic rate

Frogs:	
Small	_____
Large	_____
Rodents:	
Small	_____
Large	_____

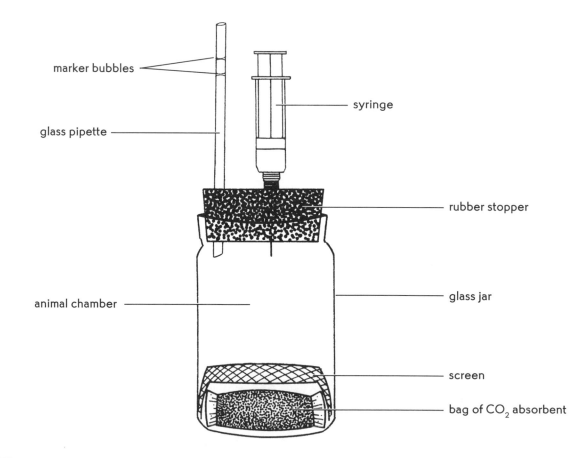

marker bubbles

glass pipette

syringe

rubber stopper

animal chamber

glass jar

screen

bag of CO_2 absorbent

FIGURE 12.1 The closed bottle respirometer to be used in this experiment.

Respiratory Quotient

The respiratory quotient (R.Q.) is the ratio of oxygen volume consumed divided by the volume of carbon dioxide produced. The R.Q. gives some indication of the nature of the substance being utilized by the animal for energy. For the complete combustion of carbohydrates the ratio is 1.0, for the combustion of lipids and proteins the ratio is 0.7. If the animal is converting carbohydrates to lipids for storage the ratio can be as high as 1.5.

Rerun the respirometer, using the same animal, but this time leave out the soda lime bag. Start with the syringe at the 2.5 ml mark because you may have to remove air. Note that the changes in air volume in the jar are not as rapid this time. The reason is that the carbon dioxide produced by the animal replaces the oxygen it consumes, thus the volume changes little.

	change per ___ min	change per hour	change per hour per gram
Frogs:			
Small	_____	_____	_____
Large	_____	_____	_____
Rodents:			
Small	_____	_____	_____
Large	_____	_____	_____

Subtract the change **without soda lime bag** from value **with soda lime bag** (oxygen consumption). Call this the difference, and note that it may be positive or negative. Subtract the difference from the oxygen consumption to get carbon dioxide production.

Run with bag – run without bag = CO₂ production

$$\text{Run with bag} - \text{run without bag} = CO_2 \text{ production}$$

Frogs:

Small _____ – _____ = _____

Large _____ – _____ = _____

Rodents:

Small _____ – _____ = _____

Large _____ – _____ = _____

R.Q. = oxygen consumption/carbon dioxide production

Frogs:

Small _____ / _____ = _____

Large _____ / _____ = _____

Rodents:

Small _____ / _____ = _____

Large _____ / _____ = _____

Control of Breathing in Man

The inspiratory center in the medulla sends messages via the splenic nerve to the diaphragm and costal muscles to cause inspiration. The inspiratory center shuts itself off by sending a message to an expiratory center in the medulla, which works by inhibiting the inspiratory center. With inhibition of the inspiratory center the muscles causing inspiration relax and passive expiration occurs.

The rate at which you breath is automatically controlled so that the tissues receive enough oxygen and so that carbon dioxide will be quickly removed. Chemoreceptors in the medulla respond directly to carbon dioxide and to the decrease in Ph caused by elevated carbon dioxide. The pH decrease is due to the formation of carbonic acid, as follows:

$$H_2O + CO_2 \rightarrow H_2CO_3 \rightarrow H^+ + HCO_3$$

As the carbon dioxide level rises, the receptors detect the carbonic acid and the inspiratory center in the medulla speeds up the breathing rate. The breathing rate is therefore primarily determined by the level of CO_2 in the blood and not the level of oxygen.

Procedure

In these experiments, **stop at once if you begin to feel faint.** If you have any medical problem with your heart or lungs, be a timekeeper and not a subject. **Work with a partner.**

a. Have your partner sit down and breathe normally. Time and record the number of breaths per minute; repeat 3 times and take the average.

b. Have the sitting partner hyperventilate by breathing as deeply and as fast as possible for 20 breaths. Then breath normally. Time and record the number of breaths over the next 2 minutes.

c. Wait until the normal breathing rate has returned. Have your partner breathe into a plastic bag for 2.5 minutes. Record the breathing rate for each half minute over the next 3 minutes. **Stop if you feel faint. Do not exceed 3 minutes.**

Normal:

Time Interval	Total breaths	Breaths per min
1.0 min	_____	_____
2.0 min	_____	_____
3.0 min	_____	_____

After hyperventilation:

Time Interval	Total breaths	Breaths per min
0.5 min	_____	_____
1.0 min	_____	_____
1.5 min	_____	_____
2.0 min	_____	_____

Breathing into the bag:

1.0 min	_____	_____
1.5 min	_____	_____
2.0 min	_____	_____
2.5 min	_____	_____
3.0 min	_____	_____

Questions:

1. Was your blood pH higher or lower than normal after hyperventilation?

2. Why did you experience a reduced urge to breathe after hyperventilation?

3. Why did your breathing rate increase after breathing into the bag?

OSMOREGULATION

Regulation of Extracellular Fluid Osmolality and Na+, K+, and Cl− Concentrations

The objective of this laboratory session is to compare the relationship between extracellular fluid osmolality and Na+, K+, and Cl− concentrations with that of the ambient water or air for a series of animal species. From these data you should be able to discuss mechanisms by which animals regulate the osmotic and ionic composition of extracellular fluids.

A number of species of euryhaline animals (those which can withstand a wide range of salinities) inhabit brackish water estuaries. The southern oyster drill, *Stramonita (Thais) haemastoma canaliculata* exhibits a

28 day LC50 (the time it takes for 50% of the individuals to die) range of 3.5 to 51.5 °/oo S (Liu, 1990). Juvenile blue crabs, *Callinectes sapidus*, exhibit 21 day LC50 values of 0 to 67 °/oo S (Guerin and Stickle, 1992). This laboratory exercise will examine the relationship between hemolymph osmolality and the osmolality of the ambient seawater in these two species stepwise adapted to 10, 20, and 30 °/oo S as well as Na+, K+, and Cl− concentrations. Previous studies on the osmotic relationship between the hemolymph and ambient seawater have been published for the Southern oyster drill (Stickle and Howey, 1975; Hildreth and Stickle, 1980) and for the Blue crab (Ballard and Abbott, 1969; Tagatz, 1971). Data on ion regulation in the blue crab has been published by Engel and Nichols (1977).

Procedure

Extracellular fluid will be collected via capillary tubes or a syringe from a variety of animal species along with a sample of the ambient water and placed into snap cap microcentrifuge tubes. Samples will be centrifuged in a Beckman Microfuge at 10,000 g for 5 min to remove cellular debris. Osmolality will be determined on the supernatant fraction with a Wescor osmometer, Na+, and K+ with a Coleman Flame Photometer, and Cl− with a Buchler-Cotlove Chloridometer. See the following pages for operating instructions for these apparatuses. Check tables in selected reprints for estimates of the ion levels in the extracellular fluid of the species with which you work. The following species and the environmental medium to which they have been acclimated are among those that may be available for your use:

Species	Medium
Callinectes sapidus (blue crab)	10, 20, 30 °/oo S
Crassostrea virginica (oyster)	10, 20, 30 °/oo S
Stramonita haemastoma (oyster drill)	10, 20, 30 °/oo S
Ligumia subrostrata (clam)	freshwater
Rangia cuneata (clam)	5, 15, °/oo S
Rana pipiens (grass frog)	freshwater
Acheta domesticus (house cricket)	terrestrial
Rattus norvegicus (lab rat)	terrestrial

Results and Conclusions

The instructor will tabulate the findings for the class. Calculate the mean and standard error of the mean of each treatment group of hemolymph samples and compare these data with ambient water and the other species data (see Appendix I for statistical discussion). Include these values in your lab report.

Discuss the osmotic and ionic response of the animals which you surveyed with respect to the external environment to which they were exposed. Use previously published literature to support or refute your observations.

Background Literature

Use these published sources to correlate your data and to help you draw relevant conclusions. They should be available in most libraries.

Ballard, B. S., Abbott, W. (1969). Osmotic accommodation in *Callinectes sapidus* Rathbun. Comp. Biochem. Physiol. 29:671–687.

Engel, D. W. and C. D. Nichols: A method for continuous *in vivo* measurement of hemolymph conductivity in crabs. J. exp. mar. Biol. Ecol. 26, 203–209 (1977).

Findley, A. M. and W. B. Stickle. 1978. Effects of salinity fluctuation on the hemolymph composition of the blue crab *Callinectes sapidus*. Marine Biology 46:9–15.

Guerin, J. L. and W. B. Stickle. 1992. Effects of salinity gradients of the tolerance and bioenergetics of juvenile blue crabs (*Callinectes sapidus*) from waters of different environmental salinities. Marine Biology 114:391–396.

Hand, S. C. and W. B. Stickle. 1978. Effects of tidal fluctuations of salinity on pericardial fluid composition of the American oyster *Crassostrea virginica*. Marine Biology 42:259–271.

Hildreth, J. and W. B. Stickle. 1980. Effects of temperature and salinity upon the osmotic composition of the southern oyster drill, *Thais haemastoma*. Biological Bulletin 159:148–161.

Liu, L. L., W. B. Stickle and E. Gnaiger. 1990. Aerobic and anoxic energy metabolism of the southern oyster drill, *Thais haemastoma,* during salinity and adaptation: a direct calorimetric study. Marine Biology 104:239–245.

Stickle, W. B. and T. W. Howey. 1975. The effects of tidal fluctuation of salinity on the hemolymph composition of the southern oyster drill *Thais haemastoma*. Marine Biology 33:309–322.

Tagatz, M. E. (1971). Osmoregulatory ability of blue crabs in different temperature-salinity combinations. Chesapeake Sci. 12:14–17.

Instructions for Wescor Vapor Pressure Osmometer

General

1. From 6 to 8 µl of sample to be tested is absorbed into a 5 mm disc of filter paper, which is then inserted into a sample chamber and sealed.

2. A thermocouple hygrometer is incorporated within the chamber. An electrical current is fed through the thermocouple, cooling it by means of the Peltier effect to a temperature below the dewpoint.

3. Water condenses from the air into the chamber and forms a thin film of liquid on the thermocouple junction. The heat of condensation raises the temperature as compared to the dewpoint temperature. The final instrument reading is proportional to the lowering of the dewpoint temperature.

4. Increasing solute concentration in the sample lowers the vapor pressure, thus the dewpoint is lowered. The dry chamber temperature is electronically compared to the dewpoint temperature. The final instrument reading is proportional to the lowering of the dewpoint temperature.

Sample Loading

1. For large sample volumes grasp the edge of a paper disc with fine tipped forceps and dip into the sample, then place it immediately in sample holder.

2. For small sample volumes place the filter paper disc in the sample holder first, then add 6–8 µl of sample via a capillary tube to the paper. The volume of the sample cannot be less than 6 µl (with this kind of paper disc) nor larger than 8 µl. Consistency in manipulation time is essential. Do not overload with sample.

3. The wet disc should adhere uniformly to the bottom of the depression in the sample holder and have a glistening appearance all over, but without a convex meniscus.

Operation

1. Gently push the sample holder into the instrument until it stops.

2. Seal the chamber by rotating the chamber sealing knob clockwise to a firm positive stop.

3. In about 2 minutes the instrument will beep. Read the digital readout as milliosmoles per kilogram (mOsm/kg).

4. Rotate chamber sealing knob about 1/4 turn counterclockwise and withdraw the sample holder slide.

5. Use a Kimwipe and wipe clean and dry out the depression on the sample holder. Repeat operation for new sample.

Calibration

1. The lab assistant will have run 290 and 1,000 mOsm/kg standards before the lab and have set the instrument readout to the standards. Therefore, do not adjust the calibration knob.

2. If calibration is necessary, it may be done under the supervision of the lab assistant.

3. With time, sloppy sample loading the thermocouple mount may become dirty. Special cleaning solutions

are required. If required, the cleaning on the thermocouple mount will be done by the instructor.

Instructions for the Coleman Flame Photometer

General

1. There are two parts to this instrument. The flame-aspirator unit, which includes the photomultiplier. The galvanometer, which has a dial reading in % transmission.

2. The gas is turned on first and ignited, then the oxygen is slowly turned on. The flow rates are preset and constant! The lab assistant will start the fire. Be sure the chimney is in place before operating the instrument.

3. The knob on the side of the flame chamber both opens the door and raises/lowers the sample holder. Note internally the wire-thin aspirating tube which sucks the sample up into the flame.

4. Use 5 ml disposable beakers to hold water blank, standards, and samples. The distinct change in the sound of the flame tells you clearly when you are aspirating or not. It takes about 30 seconds to aspirate and burn up 1 ml of fluid, so be prepared to read the galvanometer when dealing with small samples.

5. Between every sample and every standard aspirate some water and recheck the zero setting (see 1 under calibration).

6. There is a separate and different glass filter for each ion. Be sure the correct filter is in place. Recalibration is required every time you change the filter.

7. Be extremely careful about contamination of sample, standards, beakers, and diluting syringes with the salty perspiration from your hands.

Operation

1. Very precisely add the volume of water to the disposable beaker needed for dilution. Then add sample to the diluent. Be sure to swirl to mix sample and diluent.

2. The dilution rate depends on the ion concentration of the sample. The standards range from 0.1 to 0.8 mM of Na or K, therefore dilute your sample to get within this range. Check the textbook for estimated ranges of ion concentration of various animals. For example: cricket blood had about 160 mM Na and about 10 mM K, therefore:

For Na:

Add 5 µl sample to 1 ml water, giving 200X dilution.

For K:

Add 20 µl sample to 1 ml water, giving 50X dilution.

3. Check zero (the water blank) and one standard (say 0.8 mM). (See under calibration.)

4. Place disposable beaker in holder, close the door, then turn doorknob clockwise to raise holder and start aspiration.

5. Read I% transmission as soon as the needle shadow comes to rest.

6. Determine concentration from the standard curve and multiply by the dilution factor to get the concentration of the ion in the sample.

Calibration

1. While aspiration and burning distilled water (the blank), set the galvanometer meter to 0% transmission using the fine blank knob (Blk. fine) located on the flame-aspirator unit. The coarse blank knob is preset — do not alter it.

2. While aspirating and burning a standard, set the galvanometer to the % transmission corresponding to the standard (from the standard curve) by using the Galv. fine knob located on the galvanometer. The coarse Galv. knob is preset (about 1–1.5); do not alter it.

   Note: If one standard is set according to the standard curve, then all other standards will fall on the line.

3. Use the sliding galvanometer scale for final adjustment and to take care of slight variations during operation of the instrument.

4. Standard NaCl-KCl solutions in the concentration range from 0.1 to 1.0 mM have been prepared. Because of the restricted band pass (wave length) of the filter, one does not interfere with the other.

5. Periodically the aspirating tube must be reamed with a special reamer to assure continuous and even aspiration. This step will be done, when needed by the lab assistant.

Instructions for Buchler-Cotlove Chloridometer

General

1. Do not turn instrument off. Leave the instrument with a vial of blank in place.

2. The current generator has been preset (red needle) to about 13. Do not change this setting.

3. Note the click positions of the vial holder. The holder only swings out in the down position. Only use the special sized, straight-sided vials provided.

4. Always use exactly 2 ml of blank or standards, and always add 2 drops of gelatin reagent to blanks or standards. **Standards are always added to the blank.**

5. The range switch sets the concentration range (see under calibration). For terrestrial or fresh water animals use "lo" range, and for marine animals use the medium range. DO NOT touch this knob without consulting the lab assistant!

6. When the titrate switch is turned to "adjust" the stirrer and current flow starts. Be sure the vial with fluid is in place.

7. When the titrate switch is turned to "titrate," two things happen. First, the timer starts running. Second, an electrical current starts to flow which causes a release of silver ions into the sample. When all chloride ions are precipitated by the silver ions, the current flow (black) reaches the preset current flow (red) and everything is shut off.

8. Because the release of silver ions is at a constant rate, the elapsed time (in seconds) is directly related to chloride ion concentration.

9. The timer reads out in tenths of a second. **Remember to reset the timer after each run.**

10. If your sample Cl concentration is between 100–200 mM, then use low range and add 10 µl of sample of 2 ml of blank.

Operation

1. Set timer to 0 seconds. Set range to fine. Titrate switch at "titrate."

2. *Blank:* Add 2 drops of gelatin to 2 ml of blank solution (see calibration).

   Sample: Add 2 drops of gelatin to 2 ml of blank solution, then add sample.

   Standards: These were run by the lab assistant earlier, and from these a constant (k) is determined (see under calibration).

3. Place vial in a holder and raise holder until properly seated.

4. Switch titrate switch to adjust. The stirrer starts. Wait until the black needle swings near zero and stops.

5. Switch titrate switch to titrate. Wait for timer to stop (stirrer also stops), then read elapsed time.

6. Lower holder, remove vial, rinse the electrodes with a stream of water from a squirt bottle, and blot electrodes and stirrer dry with Kimwipes.

7. Reset timer, and start another run.

Calculations

1. The instrument readout is in seconds. You must have a time value for the blank and for each sample run. The lab assistant will give you the constant (k).

2. Use the following formula to get chloride ion concentration:

$$\frac{k \times \text{diff. in secs of blank and sample}}{\text{Vol of sample expressed in ml.}} \times \text{mEq Cl-/L}$$

Calibration

1. The following standards have been made:

   - Gelatin reagent: Mix dry 60:1:1 Knox unflavored gelatin #1, thymol blue, and thymol crystals. To 6.2 g of dry mix add 1 liter of hot water.

   - Blank: To 90 ml distilled water add 0.64 ml conc HNO_3 and 10 ml acetic acid.

   - Low standard: (0.5 mEq/L NaCl) Range switch to "lo." To 5 ml of 100 mEq/L NaCl (5.845 g NaCl/L) in a 1 liter volumetric add about 800 ml distilled water. Add 6.4 ml conc HNO_3 and 100 ml glacial acetic acid and then dilute to 1 liter.

   - Medium standard: (1.0 mEq/L NaCl) Range switch to "med." To 10 ml of 100 mEq/L NaCl do as for low standard.

2. The determination of the constant (k) will be done prior to the lab by the lab assistant:

$$k = \frac{\text{Vol of standard} \times \text{conc of standard}}{\text{Diff in secs of blank and standard}}$$

3. Periodically the electrodes must be polished and the electrodes trimmed. Thoroughly clean all silver polish off and blot dry. This task, when required, will be done by the lab assistant.

NITROGEN EXCRETION

Patterns of Nitrogen Excretion

The mode of nitrogen excretion used by an organism often reflects the environment in which it lives. Terrestrial animals tend to produce less toxic, more metabolically expensive forms of nitrogenous waste, whereas aquatic organisms produce "cheaper," more toxic wastes and void them in the surrounding water (Baldwin-Needham hypothesis). Some organisms, such as lungfishes and some amphibians, experience both aquatic and terrestrial environments within their lifetime. As an adaptation to this environmental variation, they have developed the ability to "switch" modes of nitrogen excretion.

The objective of this lab is to compare the patterns of nitrogen excretion throughout larval development and metamorphosis in tadpoles of the bullfrog *Rana catesbeiana*. Ammonia and urea excretion rates will be determined at several stages of development beginning with early tadpoles and ending when the frogs develop 4 legs and assume a terrestrial lifestyle.

Methods

Nitrogen excretion rates will be determined by incubating tadpoles in water baths of known volume, with water samples being collected at the end of a 1–2 hour incubation period. Water samples after incubation will be analyzed for ammonia concentration using the method of Solorzano (1969) and for urea concentration using a urease buffer (can be purchased through any major chemical supplier). Absorbance of the samples will be read after 2–24 hours.

Incubations for ammonia and urea excretion rates will be carried out at 4 or 5 stages of development over a period of several weeks, depending on how fast the tadpoles develop to metamorphosis. Thyroxin will be added to the water to hasten metamorphosis.

Urea Assay

The urea assay is based on the principle that the enzyme urease converts urea to ammonia. Therefore, two tubes will be required for determination of urea in samples (one for urea plus ammonia and one for ammonia only). A sample of 5.0 ml will be pipetted into each of the two test tubes per animal (actually four; do each sample in duplicate). Urease buffer (0.5 ml) will be pipetted into one of the tubes and both tubes will be incubated at 37°C for one hour in order for urea to be degraded to ammonia. Ammonia reagents will be added to both tubes (see next page) and then 0.5 ml urease will be added to the ammonia tubes only to account for the optical density of the urease buffer solution. Urea excretion will be determined as the difference between the total ammonia and ammonia only tubes. A standard of 0.1071 µm urea will be prepared and read. Urea excretion rate will be determined (in units of µm urea/hr) according to the following equation:

$$\frac{(\text{Sample O.D.} - \text{control O.D.}) \times \text{ml/hr}}{(\text{Standard O.D./µm urea}) \times 5 \text{ ml}}$$

Ammonia Assay

Pipette 5.0 ml of sample into a clean test tube. The sample should contain between 1 and 40 µm NH per liter. To prepare 10, 20 and 40 µm standards, pipette 5.0 ml of control water into each of three clean test tubes. Withdraw 50, 100 and 200 µl of sample water from the tubes and replace with a like volume of 1,000 µm NH/L for concentrations of 10, 20, and 40 µm NH/L. Prepare replicate test tubes for each sample, standard and control.

Add 0.2 ml phenol solution and vortex.
Add 0.2 ml nitroprusside solution and vortex.
Add 0.5 ml oxidizing solution and vortex.
Cap tubes with parafilm and place in the dark.
Read absorbance after 2–24 hours.
Calculate ammonium excretion rate (in units of µM NH/hr) according to the following equation:

$$\frac{(\text{Sample O.D.} - \text{control O.D.}) \times \text{L/hr}}{(\text{Standard O.D./µM NH/L})}$$

Background Literature

Use these published sources to substantiate or refute your experimental data. Be sure and cite the source properly in your lab report.

Jassens, P. A. 1972. The influence of ammonia on the transition to ureotelism in *Xenopus laevis*. J. Exp. Zool. 182: 357–366.

Munro, A. F. 1953. The ammonia and urea excretion of different species of Amphibia during their development and metamorphosis. Biochem. J. 54: 29–36.

Solorzano, L., 1969. Determination of ammonia in natural waters by the phenol-hypochlorite method. *Limnol. Oceanogr.*, Vol. 14, pp. 799–801.

APPENDIX 1

Basic Statistical Principles

In all biological and physical processes there is random variation. The result of this is that any experiment which we perform may give different values on repeated trials. It is the role of statistics to determine how much of this variation would be expected naturally and when a value actually is different from another. When we examine the scientific method we find that we must test our hypothesis. This step pertains not only to the performance of the testing experiment but to the proper statistical analysis of our results. In this section we will examine how to develop a proper statistical hypothesis and some ways in which we may test them.

Most statistical tests depend on data meeting certain conditions. These conditions are referred to as the **assumptions** of the test since we assume that they are met when reporting our results. The two assumptions which we will be concerned with are those of **independence** and **normality.**

Our data points must all be independent of one another. Thus, the value of one data point must in no way affect the value of another. As an example, suppose that we wanted to know the average shoe size of students in a certain class. We could not consider the measurement of the left and right shoe sizes of one individual as independent measures. An individual with a size 12 right foot is very unlikely to have a size 8 left. Measurements on different individuals are, however, independent.

The assumption of normality relates to the way a large number of data values from a given population relate to each other. This assumption is usually stated as: "the sample comes from a population with an underlying normal distribution." The normal distribution is one most of you are already familiar with, it is the "bell" shaped curve by which grades are often assigned. Roughly speaking when a large number of random data points are obtained, the largest number are near the natural center or **mean** of the population. As one moves away from this mean less and less data points are found. A mathematical equation has been developed for this distribution so that the precise probability of obtaining any point on the normal distribution may be easily determined from published tables.

We can use this normal distribution to determine the probability of obtaining the result which we observe under the conditions of our hypothesis. For this exercise we will examine the "**t-test**" to compare two treatment groups. We are assuming here that there are an equal number of observations for each of the two treatment groups.

1. First, find the means of each group. $Y_i = X_{ii}/n$; Where X_{ii} is the individual observation, and **n** is the sample size.

2. Second, find the standard deviation of each sample.

$$s = \frac{\Sigma y^2}{n}$$

The sample t is then

$$\Sigma t_s = \frac{(Y_1 - Y_2)}{1/n(s_1^2 + s_2^2)}$$

We then compare the t_s to the $t_{.05[2(n-1)]}$ obtained from a standard table of t values. If t_s is greater than the tabulated value we conclude that the two populations are different.

How to Format a Scientific Paper

Scientific papers are nearly always formatted in a precise and uniform manner. This allows researchers to efficiently synthesize material in a way that other scientists can quickly understand and, if interested, repeat the experiment. This format leaves little room for creativity, but imparts a structure that is regarded as most useful. The sequence of topics covered is as follows: **introduction, materials and methods, results,** and the **discussion**

Introduction

The opening section describes the topic of focus and summarizes any pertinent literature. Sufficient background information to allow the reader to understand and evaluate the results of the study should be presented. The introduction often offers a rationale for the present study. Choose references carefully to provide the most relevant background information rather than list an exhaustive review of the topic.

Materials and Methods

The second section specifies the exact procedure used in the study. The main purpose of this section is to provide enough detail that a competent worker can repeat the experiment. Due to the intricate nature and many details associated with this section, many readers skip this section, and proceed immediately to the results section. Careful writing of this section is critically important, however, because the cornerstone of the scientific method requires that your results be reproducible, so you must provide the exact procedure and types of equipment utilized.

Results

This is the most significant part of the paper, because it is here that you present the raw or transformed data. You should present the data in a concise, yet informative manner. Do not repeat your data in the text if they are properly presented in a Table or Figure. Use summation statistics (means, standard deviations, and standard errors) to compress your data. Do not discuss the implications of your data; save this for the discussion section. Present any graphs, tables, or figures here and cite them in the manuscript.

Discussion

This final section is often the most interesting portion of the paper. It is in the discussion that you evaluate your results and interpret them. It is here that you show the relationships among observed phenomena. Present the principles, relationships, and generalizations shown by the results. Mention any exceptions or any lack of correlation, and define unsettled points. Often, scientific controversies (an integral part of the scientific process) are generated. Relate your results to any previously published findings on the subject. Explain how and why your interpretations correlate or differ from other published work. Lastly, state your conclusion and summarize your major points.

Architectural Patterns of Animals

The basic body plans are unicellular (or acellular), cell aggregate, blind sac, and tube-within-a-tube.

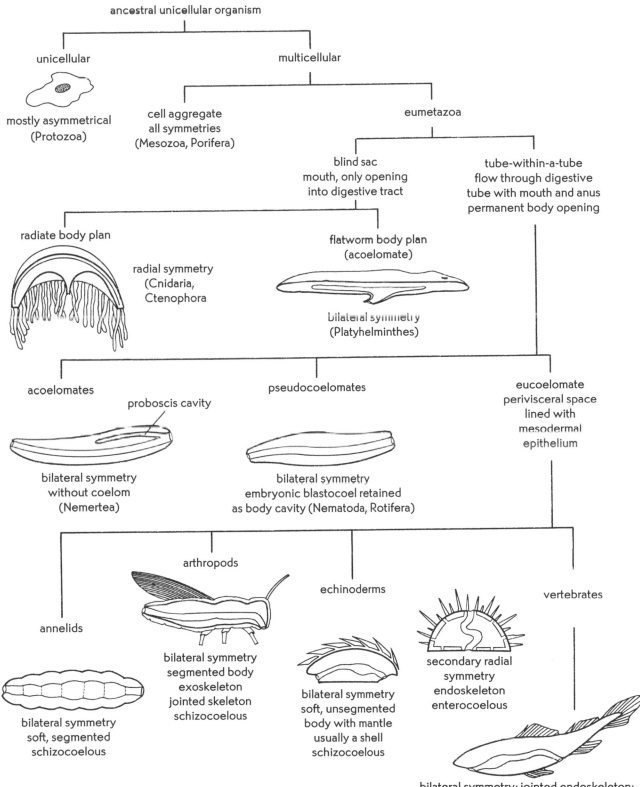

ancestral unicellular organism

unicellular

mostly asymmetrical
(Protozoa)

multicellular

cell aggregate
all symmetries
(Mesozoa, Porifera)

eumetazoa

blind sac
mouth, only opening
into digestive tract

tube-within-a-tube
flow through digestive
tube with mouth and anus
permanent body opening

radiate body plan

radial symmetry
(Cnidaria,
Ctenophora

flatworm body plan
(acoelomate)

bilateral symmetry
(Platyhelminthes)

acoelomates

proboscis cavity

bilateral symmetry
without coelom
(Nemertea)

pseudocoelomates

bilateral symmetry
embryonic blastocoel retained
as body cavity (Nematoda, Rotifera)

eucoelomate
perivisceral space
lined with
mesodermal
epithelium

arthropods

annelids

bilateral symmetry
soft, segmented
schizocoelous

bilateral symmetry
segmented body
exoskeleton
jointed skeleton
schizocoelous

echinoderms

bilateral symmetry
soft, unsegmented
body with mantle
usually a shell
schizocoelous

secondary radial
symmetry
endoskeleton
enterocoelous

vertebrates

bilateral symmetry; jointed endoskeleton;
specialized dorsal nervous system; enterocoelous

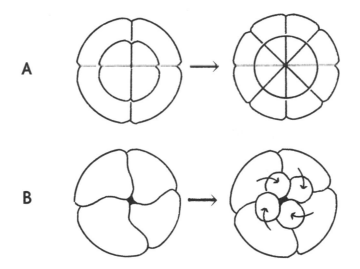

Radial and spiral cleavage. A) radial cleavage shown at eight- and 16-cell stages. B) spiral cleavage, showing the transition from four- to eight-cell stage. Arrows indicate clockwise movement of small cells (micromeres) following division from large cells (macromeres).

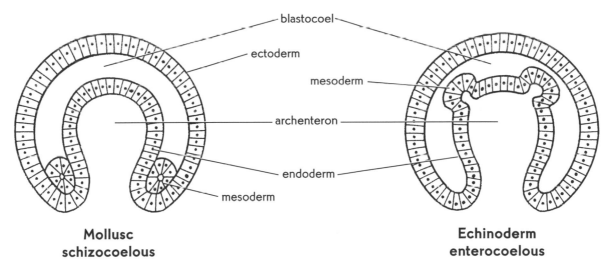

Two types of mesoderm and coelom formation: schizocoelous, in which mesoderm originates from wall of archenteron near lips of blastopore, and enterocoelous, in which mesoderm and coelom develop from endodermal pouches.

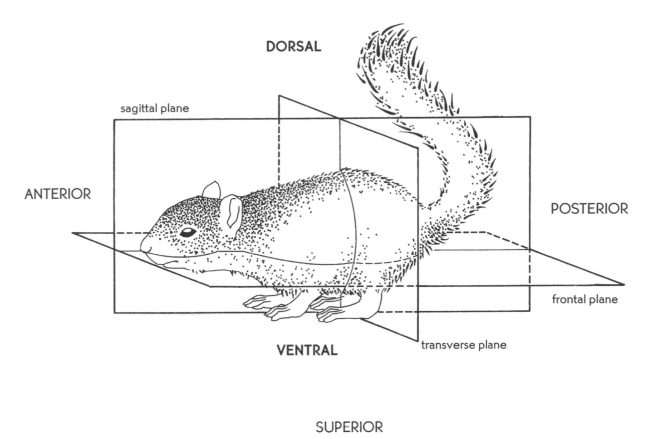

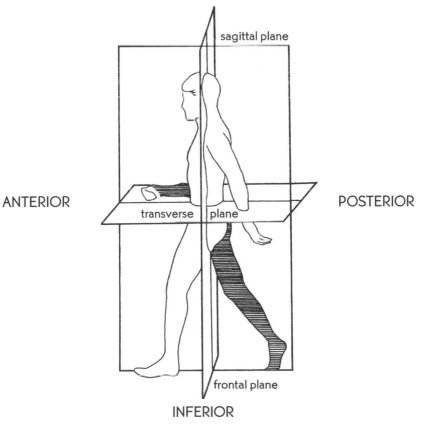

Diagrams to show the planes of the body and the differences in the terms for direction in a quadruped and in a human being.